PREFACE

In March 1987, the International Commission on Radiological Protection established a Task Group of Committee 2 to evaluate dose per unit intake for members of the public. This task covers the development of age-dependent physical and biokinetic models and the selection of appropriate biokinetic parameters.

In its first report, *ICRP Publication 56 (ICRP, 1989: Age-dependent Doses to Members of the Public from Intake of Radionuclides: Part 1)*, the Task Group gave dose coefficients for intakes by ingestion and inhalation of radioisotopes of hydrogen, carbon, strontium, zirconium, niobium, ruthenium, iodine, caesium, cerium, plutonium, americium, and neptunium.

In *ICRP Publication 67 (ICRP, 1993: Age-dependent Doses to Members of the Public from Intake of Radionuclides: Part 2 Ingestion Dose Coefficients)*, ingestion dose coefficients are given for radioisotopes of sulphur, cobalt, nickel, zinc, molybdenum, technetium, silver, tellurium, and polonium using the revised tissue weighting factors (w_T) given by the Commission *ICRP Publication 60 (ICRP, 1991: 1990 Recommendations of the International Commission on Radiological Protection)*. In addition, a generic model for the biokinetics of lead and the alkaline earth elements strontium, barium, and radium was introduced for calculating ingestion dose coefficients for radioisotopes of these elements.

Because new information on the biokinetics of plutonium and related elements in humans had become available since the issue of *ICRP Publication 56*, and information on excretion pathways is required for estimating realistic doses to the urinary bladder and colon as required by *ICRP Publication 60*, the biokinetic models used for calculating dose coefficients for radioisotopes of plutonium, americium, and neptunium were modified in *ICRP Publication 67*.

Also included in *ICRP Publication 67* are ingestion dose coefficients for the radioisotopes of the elements in *ICRP Publication 56*, using the revised w_T values and a model for systemic excretion. Generic age-dependent turnover rates for bone mineral were adopted where no specific information is available.

For this report, ingestion dose coefficients are given for radioisotopes of iron, selenium, antimony, thorium, and uranium.

The Task Group had the following membership:

Members

A. Kaul (Chairman)	J. Inaba	D. Noßke
J. C. Barton	R. W. Leggett	J. W. Stather
K. F. Eckerman	H. Métivier	

Corresponding Members

M. R. Bailey	J. D. Harrison	H. D. Roedler†
X. Chen	L. Johansson	M. Sikov
D. Crawford-Brown	G. M. Kendall	D. M. Taylor
M. Cristy	C. R. Richmond	T. Watabe

†Deceased.

The work of the Task Group was substantially supported by members of the Task Group on Dose Calculations who were also responsible for computing the dose coefficients. The membership of the Task Group at the time of preparation of the report was:

Members

K. F. Eckerman (Chairman)	K. Henrichs	I. A. Likhtarev
M. Cristy	G. M. Kendall	D. Noßke
L. T. Dillman	R. W. Leggett	

Corresponding Members

V. Berkovski	A. C. James	A. R. Reddy

During the period of preparation of this report, the membership of Committee 2 was:

1989–1993

C. B. Meinhold (Chairman)	K. F. Eckerman	N. Parmentier
W. J. Bair	A. Kaul	C. R. Richmond
A. Bouville	I. A. Likhtarev	J. W. Stather
X. Chen	O. Matsuoka	D. M. Taylor
G. Drexler	H. Métivier	R. H. Thomas

1993–1997

A. Kaul (Chairman)	F. A. Fry	M. Roy
A. Bouville	J. Inaba	J. W. Stather
X. Chen	I. A. Likhtarev	D. M. Taylor
F. T. Cross	H. Métivier	R. H. Thomas
G. Dietze	H. G. Paretzke	
K. F. Eckerman	A. R. Reddy	

GLOSSARY OF TERMS

Absorbed Dose

the physical dose quantity is given by:

$$D = \frac{d\bar{\varepsilon}}{dm}$$

where $d\bar{\varepsilon}$ is the mean energy imparted by ionising radiation to the matter in a volume element and dm is the mass of the matter in this volume element. The SI unit for absorbed dose is joule per kilogram ($J\ kg^{-1}$) and its special name is gray (Gy).

Absorbed Fraction ($AF(T \leftarrow S)_R$)

the fraction of energy emitted as a specified radiation type R in a specified source region S which is absorbed in a specified target tissue T.

Becquerel (Bq)

the name for the SI unit of activity, $1\ Bq = 1\ s^{-1}$.

Cells Near Bone Surfaces

those tissues which lie within 10 μm of endosteal surfaces, and bone surfaces lined with epithelium.

Committed Effective Dose ($E(\tau)$)

the sum of the products of the committed organ or tissue equivalent doses and the appropriate organ or tissue weighting factors (w_T), where τ is the integration time in years following the intake. The integration time is 50 y for adults and from intake to age 70 y for children.

Committed Equivalent Dose ($H_T(\tau)$)

the time integral of the equivalent dose rate in a particular tissue or organ that will be received by an individual following intake of radioactive material into the body, where τ is the integration time in years following the intake. The integration time is 50 y for adults and from intake to age 70 y for children.

Cortical Bone

equivalent to "Compact Bone" in *ICRP Publication 20* (ICRP, 1973), that is, any bone with a surface/volume ratio less than 60 $cm^2\ cm^{-3}$; in Reference Man it has a mass of 4000 g.

Dose Coefficient

committed tissue equivalent dose per unit intake at age t_0, $h_T(\tau)$ or committed effective dose per unit acute intake $e(\tau)$, where τ is the time period in years over which the dose is calculated i.e. 50 y for adults and $70 - t_0$ y for children.

Effective Dose (E)
the sum of the weighted equivalent doses in all tissues and organs of the body, given by the expression:

$$E = \sum_{T} w_T H_T$$

where H_T is the equivalent dose in tissue or organ, T, and w_T is the weighting factor for tissue T.

Equivalent Dose (H_T)
the equivalent dose, $H_{T,R}$, in tissue or organ T due to radiation R, is given by:

$$H_{T,R} = w_R D_{T,R}$$

where $D_{T,R}$ is the average absorbed dose from radiation R in tissue T and w_R is the radiation weighting factor. Since w_R is dimensionless, the units are the same as for absorbed dose, J kg^{-1}, and its special name is sievert (Sv). The total equivalent dose, H_T, is the sum of $H_{T,R}$ over all radiation types.

$$H_T = \sum_{R} H_{T,R}$$

Fractional Absorption in the Gastrointestinal Tract (f_1)
the f_1 value is the fraction of an ingested element directly absorbed to body fluids.

Gray (Gy)
the special name for the SI unit of absorbed dose.
1Gy = 1 J kg^{-1}.

Prompt Excretion
removal of activity directly from blood to urinary bladder or to the gastrointestinal tract with the clearance half-time assigned to blood.

Radiation Weighting Factor (w_R)
the radiation weighting factor is a dimensionless factor to derive the equivalent dose from the absorbed dose averaged over a tissue or organ and is based on the quality of radiation (ICRP, 1991).

Red Bone Marrow (active)
the component of marrow which contains the bulk of the haematopoietic stem cells.

Reference Man
a person with the anatomical and physiological characteristics defined in the report of the ICRP Task Group on Reference Man (ICRP, 1975).

Sievert (Sv)
the name for the SI unit of equivalent dose and effective dose.
1 Sv = 1 J kg^{-1}.

Source Region (S)
region within the body containing the radionuclide. The region may be an organ, a tissue, or the contents of the gastrointestinal tract or urinary bladder, or the surfaces of tissues such as in the skeleton.

Specific Effective Energy ($SEE(T \leftarrow S)_R$)
the energy, suitably modified for radiation weighting factor, imparted per unit mass of a target tissue, T, as a consequence of the emission of a specified radiation, R, from a transformation occurring in source region S expressed as Sv $(Bq\ s)^{-1}$.

Target Tissue
tissue or organ in which radiation is absorbed.

Tissue Weighting Factor (w_T)
the factor by which the equivalent dose in a tissue or organ is weighted to represent the relative contributions of that tissue or organ to the total detriment resulting from uniform irradiation of the body (ICRP, 1991).

Trabecular Bone
equivalent to "Cancellous Bone" in *ICRP Publication 20* (ICRP, 1973), i.e. any bone with a surface/volume ratio greater than 60 $cm^2\ cm^{-3}$; in Reference Man it has a mass of 1000 g.

Transfer Compartment
the compartment introduced (for mathematical convenience) into most of the biokinetic models used in this report to account for the translocation of the radioactive material through the body fluids from where they are deposited in tissues.

References

ICRP (1973). *Alkaline Earth Metabolism in Adult Man.* ICRP Publication 20, Pergamon Press, Oxford.
ICRP (1975). *Report of the Task Group on Reference Man.* ICRP Publication 23, Pergamon Press, Oxford.
ICRP (1991). *Recommendations of the International Commission on Radiological Protection.* ICRP Publication 60. *Annals of the ICRP.* **21**(1–3), Pergamon Press, Oxford.

1. INTRODUCTION

(1) The present report on age-dependent dose coefficients to members of the public from intakes by ingestion of radioisotopes of the following elements: iron, selenium, antimony, thorium, and uranium follows *ICRP Publications 56* (ICRP, 1989) and *67* (ICRP, 1993) which also gave age-dependent dose coefficients. This report gives parameters for the tissue distribution and retention of these elements together with data on urinary and faecal excretion. Information on excretion pathways is required because of the need to estimate realistic doses to the urinary bladder and colon (ICRP, 1991a).

(2) Dose coefficients have been calculated for radioisotopes of the elements dealt with in this report which are expected to be released into the environment as a result of human activities and considered to be of significance for environmental radiation protection purposes.

(3) The age-dependent ingestion dose coefficients for selected radioisotopes of the elements in this report are based on reviews of biokinetic data in the literature and the general biokinetic and dosimetric models described in *ICRP Publications 56* and *67*.

(4) The generic model structure for plutonium, americium and neptunium given in *ICRP Publication 67* (ICRP, 1993) has been applied to thorium and the generic model structure for the alkaline earths given in *ICRP Publication 67* has been applied to uranium. The movement of elements between body compartments in a recycling model is best described by a series of transfer rates. The parameters are derived from model fits to experimental data and are given to a sufficient precision to maintain numerical accuracy.

(5) Where no clear evidence on age dependence of organ distribution and retention appeared to be available, the biokinetic data for adults were adopted for infants and children. This assumption was made in *ICRP Publications 56* and *67*, and is usually expected to lead to an overestimate of the dose coefficients.

(6) If no relevant biokinetic data were found for humans, appropriate data were based on animal experiments as far as possible.

(7) Many elements incorporated into food may be more readily absorbed from the gastrointestinal tract than inorganic forms of these elements (ICRP, 1989). This is taken into account in the choice of the recommended f_1 value. When no specific information on absorption from food is available, default values are adopted (ICRP, 1993).

(8) Absorption of radionuclides tends to be greatest in the newborn but the results of animal studies suggest that uptake from the gut progressively decreases with age, approaching adult values by about the time of weaning in most cases. As described in *ICRP Publication 56*, where no human or animal data are available, the following approach, based on the general body of animal data supported by limited human data, was adopted for 3-mo-old infants. For elements with a fractional absorption in the adult of 0.001 or less, a value of 10 times that for the adult is assumed. For fractional absorption values between 0.01 and 0.5 in the adult, an increase by a factor of 2 is assumed. For the elements covered in this report, none have f_1 values for adults between 0.01 and 0.001. This general approach is based on the recommendations of an NEA Expert Group (NEA/OECD, 1988). For the group of elements with an assumed increase of the f_1 value of a factor 10 the actinides are examples for which the experimental evidence is well established (ICRP, 1986, 1989). For the group of elements with an assumed increase in the f_1 value of a factor of 2, niobium is an example which is

supported by data from rats and guinea pigs (ICRP, 1989). For absorption values greater than 0.5 in the adult, the approach adopted here is to assume complete absorption by 3-mo-old infants ($f_1 = 1$).

(9) Absorption of material from the GI tract is assumed to take place in the small intestine. If complete absorption is indicated in the biokinetic model then for computational reasons a fraction 0.99 of the ingested material is taken to be absorbed to the transfer compartment as soon as it enters the small intestine; a fraction 0.01 of the ingested material enters the lower segments of the GI tract.

(10) For loss of activity from the transfer compartment, a half-time of 0.25 d has been adopted (ICRP, 1979), unless specified otherwise.

(11) The tissue weighting factors, w_T used for calculating the effective dose coefficients are those given by ICRP in its 1990 recommendations (ICRP, 1991) (Table 1). Dose coefficients are given in this report for both the ovaries and testes, the higher of which is applied to the gonad weighting factor for the calculation of effective dose. The Commission has clarified application of the colon weighting factor given in Table 1. The weighting factor is to be applied to the mass average of the equivalent dose in the walls of the upper and lower large intestine of the gastrointestinal tract (*ICRP Publication 30, 1979*). The upper large intestine is no longer included in the remainder tissues. Age-specific masses of the walls of the gastrointestinal tract are given in Table 1.-1 of *ICRP Publication 56.* Since the relative masses of the walls of the upper large intestine (ULI) and lower large intestine (LLI) are independent of age the equivalent dose to the colon H_{colon} is given as:

$$H_{colon} = 0.57\ H_{ULI} + 0.43\ H_{LLI}$$

where H_{ULI} and H_{LLI} are the equivalent doses in the walls of the ULI and LLI, respectively.

(12) In the 1990 recommendations (ICRP, 1991), the urinary bladder is given an explicit w_T value as shown in Table 1. Consequently, a urinary bladder model is formulated for calculating the additional equivalent dose to the bladder wall from activity in the urine. The radiation dose to the bladder wall involves a complex relationship between urine flow rate, voiding, and urine volume present in the bladder when the radionuclide entered the bladder and is critically dependent on the geometrical relationship between the wall of the bladder and its contents. For radiation protection purposes, such a model was developed by Snyder and Ford (1976) and was extended by Smith *et al.* (1982) to administered radio-pharmaceuticals. This model was adopted in *ICRP Publication 53* (ICRP, 1987) and modified in *ICRP Publication 67* to facilitate its application to long-lived radionuclides. The bladder is taken to be of fixed size containing 15, 25, 65, 75, 85, and 115 ml of urine in 3-mo-old infant, 1-, 5-, 10-, 15-y-old children, and adults, respectively. These volumes represent the average content of the bladder during the time period between voids. The rate at which radionuclides enter the bladder is based on their elimination rates from body tissues and the urinary to faecal excretion ratio given with the biokinetic data. For some elements, the biokinetic data directly address excretion, as in the biokinetic models for the alkaline earth and actinide elements which are applied to uranium and thorium respectively in this report. While the urinary to faecal excretion ratio for an element may vary with time, a constant value is judged to be adequate for present dosimetric purposes. It is not intended that the stated ratios should be used for the interpretation of bioassay measurements where temporal variations in the ratio may be important. The number of voids per day for the 3-mo-old and 1-y-old are taken as 20 and 16, respectively (Goellner *et al.*, 1981). For all other ages

Table 1. Tissue weighting factors[1]

Tissue or organ	Tissue weighting factor, w_T
Gonads[2]	0.20
Bone marrow (red)	0.12
Colon[2]	0.12
Lung	0.12
Stomach	0.12
Bladder	0.05
Breast	0.05
Liver	0.05
Oesophagus[3]	0.05
Thyroid	0.05
Skin	0.01
Bone surface	0.01
Remainder[4-6]	0.05

[1] The values have been developed from a reference population of equal numbers of both sexes and a wide range of ages. In the definition of effective dose they apply to workers, to the whole population, and to either sex.

[2] See Paragraph (11).

[3] See Paragraph (13).

[4] For purposes of calculation, the remainder is composed of the following nine additional tissues and organs: adrenals, brain, small intestine, kidney, muscle, pancreas, spleen, thymus and uterus. The list includes organs which are likely to be selectively irradiated. Some organs in the list are known to be susceptible to cancer induction. If other tissues and organs subsequently become identified as having a significant risk of induced cancer they will then be included either with a specific w_T or in this additional list constituting the remainder. The latter may also include other tissues or organs selectively irradiated.

[5] In the exceptional case in which the most exposed remainder tissue or organ receives the highest committed equivalent dose of all organs, a weighting factor of 0.025 (half of remainder) is applied to that tissue or organ and 0.025 (half of remainder) to the mass average committed equivalent dose in the rest of the remainder tissues and organs (ICRP, 1993).

[6] The new model for the respiratory tract issued as *ICRP Publication 66* (1994) includes the extrathoracic, ET, region which has been defined as an additional remainder tissue. The ET region is not listed in the dose tables given in this report since it is never of practical importance for ingestion of the nuclides considered here.

6 voids per d are assumed (Larsson and Victor, 1988; Bloom *et al.*, 1993). The volumes of the bladder contents, noted above, correspond to one-half of the volume of urine excreted in a single voiding (ICRP, 1975). To represent the kinetics of the bladder in terms of first order processes, the rate of elimination from the bladder is taken to be twice the number of voids per day. That is, the elimination rate from the bladder would be equivalent to 40, 32, 12, 12, 12, and 12 d^{-1} for the 3-mo-old infants, 1-, 5-, 10-, 15-y-old children, and the adults, respectively.

(13) The Commission has assigned an explicit w_T to the oesophagus (Table 1). The model for the gastrointestinal tract adopted here is that used in *ICRP Publication 30*

(ICRP, 1979) and contains no information on the movement of radionuclides through the oesophagus. Since the transit of materials through the oesophagus is rapid, only the dose from penetrating radiation emitted from other source regions needs to be considered. In the absence of a dosimetric model for the oesophagus, the specific absorbed fraction for thymus is used as a surrogate (ICRP, 1991).

(14) The activity in the upper and lower large intestine at times following the intake of a radionuclide includes activity being eliminated from the systemic circulation. Except where otherwise indicated, it is assumed that systemic activity lost in faeces is secreted into the upper large intestine. The dose to the lung from ingested radionuclides is computed as the average dose to the lung tissues rather than a composite of the doses to cells at risk. This was also the approach used in *ICRP Publications 56* and *67*.

(15) Dose coefficients for ingestion by 3-mo-old infants, 1-, 5-, 10- and 15-y-old children, and 20-y-old adults are given in this report. Exceptions are made for uranium and thorium for which the transfer rates for the adult apply to ages ≥25 y. This approach is used because some of the transfer rates in the biokinetic models for these elements are equated with bone formation rates, which are expected to remain elevated during the early part of the third decade of life. In the calculations of the activity in source regions of the body following intakes at the different ages, continuous changes with age in the transfer rates governing its distribution and retention are obtained by linear interpolation according to age. This also applies to the transfer of activity from the small intestine to body fluids. For application to other ages the Task Group considered that tissue doses can be estimated by applying the age-specific dose coefficients to the age ranges given below:

3 mo: infants from 0 to 12 mo of age
1 y: from 1 y to 2 y
5 y: more than 2 to 7 y
10 y: more than 7 to 12 y
15 y: more than 12 to 17 y
adult: more than 17 y.

As in *ICRP Publications 56* and *67*, a single reference subject is used to represent each age group. Generally, parameter values for males have been adopted because of the availability of biokinetic data. Where there are known gender differences in the biokinetics of an element this is noted in the relevant section of the biokinetic data.

(16) The dose coefficients calculated in this report and in *ICRP Publications 56* and *67* are for acute intakes. For chronic intakes, doses per unit intake could be somewhat less than those calculated here where growth is significant during the period of intake. However, since the age ranges were selected to account for significant changes in growth and biokinetics during life, these coefficients can also be applied to chronic intakes for protection purposes by determining the committed dose for each year's intake and summing for intakes over all years.

References

Bloom, D. A., Seeley, W. W., Ritchey, M. L. and McGuire, E. J. (1993). Toilet habits and continence in children: An opportunity sampling in search of normal parameters. *J. Urol.* **149**, 1087–1090.

Goellner, M. H., Ziegler, E. E. and Forman, S. J. (1981). Urination during the first three years of life. *Nephron* **28**, 174–178.

ICRP (1975). *Report of the Task Group on Reference Man*, ICRP Publication 23. Pergamon Press, Oxford.

ICRP (1979). *Limits for Intakes of Radionuclides by Workers*, ICRP Publication 30, Part 1. *Annals of the ICRP* **2**(3/4). Pergamon Press, Oxford.

ICRP (1986). *The Metabolism of Plutonium and Related Elements*, ICRP Publication 48. *Annals of the ICRP* **16**(2/3). Pergamon Press, Oxford.

ICRP (1987). *Radiation Dose to Patients from Radiopharmaceuticals*, ICRP Publication 53. *Annals of the ICRP* **18**(1–4). Pergamon Press, Oxford.

ICRP (1989). *Age-dependent Doses to Members of the Public from Intake of Radionuclides: Part 1*, ICRP Publication 56. *Annals of the ICRP* **20**(2). Pergamon Press, Oxford.

ICRP (1991). *1990 Recommendations of the International Commission on Radiological Protection*, ICRP Publication 60. *Annals of the ICRP* **21**(1–3). Pergamon Press, Oxford.

ICRP (1993). *Age-dependent Doses to Members of the Public from Intake of Radionuclides: Part 2 Ingestion Dose Coefficients*, ICRP Publication 67. *Annals of the ICRP* **23**(3/4). Elsevier Science Ltd, Oxford.

ICRP (1994). *Human Respiratory Tract Model for Radiological Protection.* ICRP Publication 66. *Annals of the ICRP* **24**(1–4). Elsevier Science Ltd, Oxford.

Larsson, G. and Victor, A. (1988). Micturition patterns in a healthy female population, studied with a frequency/volume chart. *J. Urol. Nephrol.* **114** (Suppl.), 53–57.

NEA/OECD (1988). *Committee on Radiation Protection and Public Health. Report of an Expert Group on Gut Transfer Factors.* NEA/OECD Report, Paris.

Smith, T., Veall, N. and Wootten, R. (1982). Bladder wall dose from administered radiopharmaceuticals: The effect on variation in urine flow rate, voiding interval and initial bladder content. *Radiat. Prot. Dosim.* **2**, 183–189.

Snyder, W. S. and Ford, M. R. (1976). Estimation of doses to the urinary bladder and to the gonads. In: *Radiopharmaceutical Dosimetry Symposium* (Proc. Conf. Oak Ridge, Tennessee, April 1976). HEW Publication (FDA 76-8044), pp. 313–349. Department of Health, Education and Welfare, Bureau of Radiological Health, Rockville, MD.

COMPUTATION OF AGE-DEPENDENT EFFECTIVE DOSE COEFFICIENTS

(17) Several subsidiary dosimetric quantities have proved useful for the dosimetry of incorporated radionuclides. Following an intake to the body of a radioactive material, there is a period during which the material gives rise to equivalent doses in the tissues of the body at varying rates. The time integral of the equivalent dose rate is called the committed equivalent dose, $H_T(\tau)$ where τ is the integration time (in years) following intake. If τ is not specified, it is implied that the value is 50 y for adults and from intake to age 70 y for infants and children. By extension, the committed effective dose, $E(\tau)$ is similarly defined.

(18) *ICRP Publication 56* (ICRP, 1989) outlined the computational formulations of the age-dependent dosimetric quantities. Some additional discussion is necessary in implementing, in an age-dependent manner, the effective dose quantity defined in the 1990 recommendations (ICRP, 1991a).

(19) Following *ICRP Publication 56*, the equivalent dose rate at age t in target organ or tissue T due to an acute intake of a radionuclide by an individual of age t_0 at the time of intake, $\dot{H}_T(t, t_0)$, is expressed as:

$$\dot{H}_T(t, t_0) = \boldsymbol{C} \sum_{S} q_S(t, t_0) \mathrm{SEE}(\mathrm{T} \leftarrow \mathrm{S}; t)$$

where $q_S(t, t_0)$ is the activity of the radionuclide in source region S at age t after intake at age t_0 and SEE(T←S; t) is the specific effective energy deposited in target organ T per nuclear transformation in source region S at age t, and $\boldsymbol{C}$ is any numerical constant required to reconcile the units of q and SEE. The total committed equivalent dose in T accumulated by age 70 y due to the intake at age t_0, $H_T(70 - t_0)$, is:

$$H_T(70 - t_0) = \int_{t_0}^{70} \dot{H}_T(t, t_0)\, \mathrm{d}t = \boldsymbol{C} \int_{t_0}^{70} \sum_{S} q_S(t, t_0) \mathrm{SEE}(\mathrm{T} \leftarrow \mathrm{S}; t)\, \mathrm{d}t$$

When a radionuclide decays to a decay product which is itself radioactive an exactly similar set of equations applies to doses from the decay product(s).

(20) The committed effective dose accumulated by age 70 y due to the intake at age t_0, $E(70 - t_0)$, is:

$$E(70 - t_0) = \sum_{T=1}^{12} w_T H_T(70 - t_0) + w_{\mathrm{remainder}} H_{\mathrm{remainder}}(70 - t_0)$$

where the summation extends over the twelve organs and tissues in Table 1 assigned explicit weighting factors, $w_{\mathrm{remainder}}$ is the weighting factor for the remainder tissues, and $H_{\mathrm{remainder}}$ $(70 - t_0)$ is the committed equivalent dose accumulated by age 70 y in remainder tissues. The treatment of the remainder dose is explained below. The equivalent dose in the remainder, $H_{\mathrm{remainder}}(70 - t_0)$, is formulated in terms of the average equivalent dose in the tissues comprising the remainder (ICRP, 1991a). For the adult, $H_{\mathrm{remainder}}(70 - t_0)$ can be evaluated directly from the dose equivalent in these tissues since their masses are independent of age; see Paragraphs (6) and (7) of *ICRP Publication 61* (ICRP, 1991b). For children, however, the evaluation of $H_{\mathrm{remainder}}(70 - t_0)$ must reflect the age-dependence of the organ masses.

(21) The computational form of the equivalent dose rate in the remainder depends upon whether any organ or tissue of the remainder experiences a committed equivalent dose in excess of the committed equivalent dose in organs and tissues given explicit weighting factors (ICRP, 1991a,b). Let H_{max} denote the maximum committed equivalent dose among all organs with explicit weighting factors, and $H_{T'}$ the maximum committed equivalent dose among organs of the remainder. The equivalent dose rate in the remainder tissues, $\dot{H}_{remainder}(t, t_0)$, is then

$$\dot{H}_{remainder}(t, t_0) = \begin{cases} \dfrac{\sum_{T=1}^{T=9} m_T(t)\dot{H}_T(t, t_0)}{\sum_{T=1}^{9} m_T(t)}, & \text{if } H_{T'} \leq H_{max} \\ 0.5\left[\dfrac{\sum_{T=1(T\neq T')}^{T=9} m_T(t)\dot{H}_T(t, t_0)}{\sum_{T=1(T\neq T')}^{T=9} m_T(t)} + \dot{H}_{T'}(t, t_0)\right], & \text{if } H_{T'} > H_{max} \end{cases}$$

where the summation extends over the nine individual organs and tissues of the remainder. In implementing the above equation, the organ masses, $m_T(t)$, are obtained by linear interpolation with time of the data in Table 1.-1 of *ICRP Publication 56* (ICRP, 1989). The committed equivalent dose accumulated by age 70 y in the remainder tissue due to an intake at age t_0 is

$$H_{remainder}(70 - t_0) = \int_{t_0}^{70} \dot{H}_{remainder}(t, t_0)\mathrm{d}t$$

(22) The dose coefficient is the committed tissue equivalent dose per unit intake at age t_0, $h_T(\tau)$, or committed effective dose per unit intake, $e(\tau)$, where τ is the time period in years over which the dose is calculated, i.e. 50 y for adults and $70 - t_0$ y for children. Thus $h_T(\tau)$ and $e(\tau)$ in Sv Bq^{-1} correspond respectively to $H_T(\tau)$ and $E(\tau)$ (given for children in Paragraphs 19 and 20) resulting from the intake of 1 Bq.

References

ICRP (1989). *Age-dependent Doses to Members of the Public from Intake of Radionuclides*, ICRP Publication 56, Part 1. *Annals of the ICRP* **20**(2). Pergamon Press, Oxford.

ICRP (1991a). *1990 Recommendations of the International Commission on Radiological Protection*, ICRP Publication 60. *Annals of the ICRP* **21**(1-3). Pergamon Press, Oxford.

ICRP (1991b). *Annual Limits on Intake of Radionuclides by Workers Based on the 1990 Recommendations*, ICRP Publication 61. *Annals of the ICRP* **21**(4). Pergamon Press, Oxford.

PREFACE

(23) Based upon a review of biokinetic data for selected elements, age-dependent reference values of uptake, distribution and excretion have been adopted for calculating dose coefficients for different ages at intake. In the present report the Task Group has calculated ingestion dose coefficients for the following radionuclides:

^{55}Fe, ^{59}Fe, ^{75}Se, ^{79}Se, ^{124}Sb, ^{125}Sb, ^{126}Sb, ^{228}Th, ^{230}Th, ^{232}Th, ^{234}Th, ^{232}U, ^{233}U, ^{234}U, ^{235}U, ^{236}U, ^{238}U.

1. IRON

Uptake to Blood

(24) *Adults.* Because of the nutritional significance of iron (Fe), its biokinetics and metabolism in man and animals have been investigated extensively (ICRP, 1975; Bothwell *et al.*, 1979; NEA/OECD, 1988). Inorganic iron, usually as ferric or ferrous salts, is present in many foods as a natural constituent or after supplementation with iron, and as the major form of ambient iron present in water and air. Biologically incorporated iron, usually in haem from haemoglobin, myoglobin, and cytochromes, is derived from animal tissues and/or their constituents present in the diet.

(25) In normal iron balance, iron absorption occurs predominantly in the small intestine and is regulated such that absorption replaces loss. Iron absorption is influenced by six major factors: (1) the amount of iron in the diet; (2) age; (3) gender; (4) the body's state of iron repletion; (5) the chemical form ingested; and (6) substances in the diet and gastrointestinal secretions which act to alter iron absorption (Bothwell *et al.*, 1979; NEA/OECD, 1988). Iron deficiency is a normal feature of rapid growth, menstruation, pregnancy, lactation, and/or variations in diet. Iron incorporated in food of animal origin is generally better absorbed than that from vegetables (Layrisse and Martinez-Torres, 1971; Martinez-Torres and Layrisse, 1973). In normal subjects, f_1 values of 0.01–0.07 have been obtained when iron was ingested with a wide variety of vegetable foods, whereas f_1 values of 0.1–0.2 are typically obtained when iron is added to meat and fish (Bothwell *et al.*, 1979). Likewise, haem iron is usually better absorbed than non-haem iron in normal and iron-deficient subjects (Layrisse and Martinez-Torres, 1972; Bjorn-Rasmussen *et al.*, 1974). In the gastrointestinal tract, ferrous iron tends to be absorbed more efficiently than ferric because of its greater solubility (NEA/OECD, 1988). Intraliminal factors such as gastric hydrochloric acid, bile, and certain organic and amino acids can augment iron absorption, whereas bicarbonate from pancreatic secretions, phosphates, phytates, carbonates, tannates, oxalates, and/or EDTA can decrease iron absorption. Healthy older adults appear to absorb iron similarly to younger adults (Freiman *et al.*, 1963; Marx, 1979).

(26) The f_1 value of 0.1 for iron previously recommended for workers in *ICRP Publication 30* (ICRP, 1980) approximates iron absorption data obtained in adult males with normal iron status. This value adequately represents iron absorption in many cases, e.g. in healthy adult male and postmenopausal female subjects, with iron in inorganic forms, and in vegetarian diets. Iron absorption measured in pregnant women gave f_1 values of about 0.1, 0.25, and 0.3 in the first, second, and third trimester. The iron requirements for menstruating or lactating females, based on losses of menstrual blood and milk, may be somewhat higher than in normal adult males although iron absorption does not appear to have been measured directly in these circumstances (Bothwell *et al.*, 1979; NEA/OECD, 1988). For the dose coefficients given in this report an f_1 value of 0.1 has been adopted.

(27) *Children.* Gorten *et al.* (1963) reported a fractional absorption of ^{59}Fe of 0.32 (range 0.068–0.74) following administration to healthy premature infants (1–2-wk-old) as ascorbate in a milk meal. In infants younger than 1.5 mo, values of fractional absorption of 0.56–0.91 were obtained using ^{59}Fe citrate administered in a milk meal; in infants 1.5–3 mo of age, f_1 values obtained were 0.15–0.38 (Garby and Sjolin, 1959). Children in

the first year of life had a fractional absorption of 0.48–0.7 of iron administered with breast milk (Saarinen *et al.*, 1977), 0.03–0.3 of iron administered with infant formulas based on soya protein extract or cow's milk (Rios *et al.*, 1975; Saarinen and Siimes, 1977), and about 0.03 from carrier-free radioiron activity (Rios *et al.*, 1975; Saarinen and Siimes, 1977; Saarinen *et al.*, 1977). Saarinen *et al.* (1977) also compared the absorption of ^{59}Fe in 6–7-mo-old infants who had been exclusively breast-fed or had been weaned from breast to cow's milk prior to the age of 2 mo. In each case, ^{59}Fe sulphate was administered during feeding after a 3 h fast and the absorption values obtained were 0.5 ± 0.08 (SEM, $n = 11$) for the breast-fed infants and 0.2 ± 0.05 ($n = 16$) for those fed cow's milk. Iron absorption in infants and children is inversely related to age but has not usually been measured in direct comparison with that in adults (Cristy and Leggett, 1986). In one comparative study, children 4–52-mo-old absorbed approximately 0.1 of the ^{59}Fe from a milk meal, in comparison with adult males who absorbed 0.028 from a similar meal (Schulz and Smith, 1958). The absorption of ^{59}Fe biologically incorporated in eggs was approximately 0.11 in children 1–4.5 y of age and 0.06 in children 5–15 y of age (Schulz and Smith, 1958). ^{59}Fe-labelled ferrous ascorbate administered in lemonade yielded mean absorption values of 0.08–0.16 in children 7–8 y of age, and 0.15–0.17 in children 9–10 y of age (Darby *et al.*, 1947). The higher values obtained in the older children in this experiment could be related to increased iron requirements during growth and development. After a 12 h fast, four normal children ages 6–11 y were given ^{59}Fe with 5 mg of carrier iron, both in the form of ferrous sulphate. The percentage absorption values were 5, 8, 17, and 27% (mean 13.5%) (Erlandson *et al.*, 1962). In adolescence enhanced absorption of iron would also be expected; the daily requirement of iron in adolescent males and females is respectively about 30 and 50% higher than in adults (Bothwell *et al.*, 1979).

(28) Although the published values of absorption are quite variable, those for children are generally higher than those for adults, attributable in part to the relatively greater requirements of iron for growth and development. For all forms of iron f_1 values of 0.6 for 3-mo-old infants, and 0.2 for 1-, 5-, 10-, and 15-y-old children are adopted here.

Distribution and retention

(29) The biokinetics of iron has been extensively studied (ICRP, 1975; Bothwell *et al.*, 1979; NEA/OECD, 1988). There are three pathways that significantly determine the distribution of iron in the body. The most important is a pathway that starts with the uptake of transferrin iron by the erythroid marrow for incorporation into haemoglobin. Most of this iron subsequently appears in circulating red blood cells, where it remains for the life of the cell (about 120 d in the adult) and, after degradation of the haemoglobin, enters the reticuloendothelial system from which it may return to plasma. A fraction of transferrin-bound iron rapidly enters the reticuloendothelial system. This fraction is derived from extruded components of developing cells or from defective cells that do not enter the circulation, or which are phagocytosed as soon as they do. Such iron may be referred to as the *wastage iron of erythropoiesis*. However, most iron may be the circulation is fixed in the red blood cells.

(30) The second of the three pathways is that followed by the transferrin-bound iron which leaves the plasma for the extravascular spaces and returns to the plasma via the lymphatic system. The third pathway is taken by the iron that is delivered to the parenchymal tissues, mainly the liver, the major depot of iron after the marrow.

(31) The total amount of iron in the adult male is about 3.5 g. The major iron constituents can be grouped into two categories. The first consists of compounds which fulfil well-defined physiological functions and are referred to as *essential iron.* The second category is *storage iron* because its major role involves the regulation of iron homeostasis and the maintenance of an iron reserve to assure an adequate supply of iron for production of essential iron compounds.

(32) The essential iron compounds include haemoglobin, whose function is the transport of oxygen from the lungs to tissue, and mitochondrial iron proteins, which are essential for the oxidative production of cellular energy. Haemoglobin accounts for over two-thirds of the iron in the body and almost all of the iron in blood. Mitochondrial iron represents about 3% of the total body iron. Iron present in myoglobin, primarily associated with muscle, is about 10% of the total body iron.

(33) Storage iron compounds are present primarily in the liver, reticuloendothelial cells, and the erythroid precursors of the bone marrow. The amount of storage iron can vary over a wide range without an apparent impairment of body function. In the adult, about 1 g of iron storage is assumed.

Structure of the biokinetic model for iron

(34) The retention model for iron given in *ICRP Publication 53* (ICRP, 1987) was based on the model in MIRD Dose Report No. 11 (Robertson *et al.*, 1983). That model has been criticised because it did not consider the small iron losses from the body (Johnson and Dunford, 1984, 1985). *ICRP Publication 53* did include age-dependent dose coefficients, however they were based on the adult biokinetics, allowing only for changes in body mass. Because of these limitations a more detailed iron model structure has been formulated to incorporate information on age-specific kinetics and to address iron losses from the body. The structure of the model is given in Fig. 1.1.

(35) In this model, blood is represented by plasma and red blood cells (RBCs). The soft tissue pool labelled "rapid turnover" includes extravascular iron which exchanges rapidly with the plasma. The red blood cell compartment contains iron incorporated into haemoglobin by the erythroid marrow in the red marrow kinetic compartment labelled "synthesis". The destruction of RBCs is viewed as occurring in the red marrow pool labelled "transit". The destruction of red blood cells and the return of haemoglobin iron to plasma may occur at other locations in the body (e.g. within the spleen), but for the purposes of a kinetic model it is convenient to consider these processes to occur within a single pool.

(36) Storage iron is included in the liver, spleen, and red marrow pool labelled "storage". The soft tissue pool labelled "storage" includes both iron incorporated in myoglobin and additional storage iron of the reticuloendothelial system not associated with the red marrow, liver, and spleen. Liver is viewed as consisting of two pools: Liver 1 (transit) consists of parenchymal tissues that exchange iron with plasma, while Liver 2 (storage) is associated with the reticuloendothelial system.

(37) Iron loss from the body is due, to a large extent, to the exfoliation of cells from its surfaces. With the physiological turnover of the epithelium of the skin, the gastrointestinal tract and the genito-urinary tract, a small amount of endogenous iron is lost. There is a much greater loss of iron when red blood cells are lost than when epithelial cells are shed. In males, postmenopausal females and prepubertal girls, about two-thirds of the total iron loss is from the gastrointestinal tract. Most of the remaining losses occur from the skin, and only a tiny fraction in the urine. Losses through the

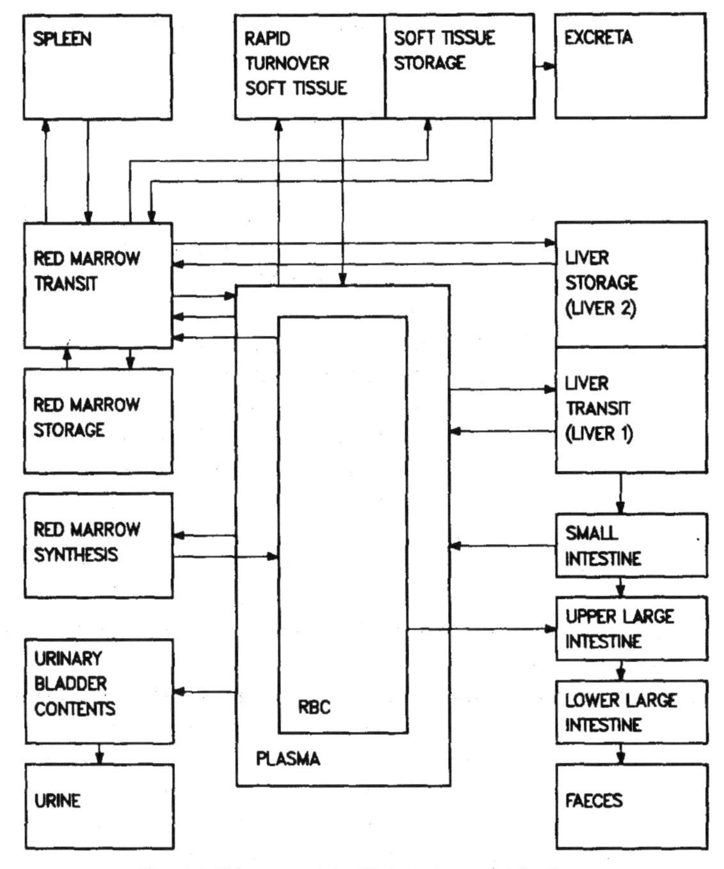

Fig. 1.1. Diagram of the biokinetic model for iron.

gastrointestinal tract reflect lost of epithelium cells, blood normally present in faeces, and biliary excretion. In the model, the biliary excretion is from Liver 1 (transit) into the small intestine from which it may return to plasma while blood loss is taken to enter the upper large intestine. Losses from the skin are assumed to be from the soft tissue storage compartment.

Parameter values for adults

(38) Parameter values assigned to the adult are based in part on the kinetics of radioiron as observed in the clinical setting and summarised by MIRD Dose Report No. 11 (Robertson *et al.*, 1983). In addition, the extensive summary of iron metabolism by Bothwell and co-workers (1979) provides considerable information on the status of the iron pools in man, particularly the adult male. In choosing model parameters it is important to select values that are appropriate for "normal iron status" and that the values be consistent with conditions of chronic exposure, rather than the kinetics of an acute intake or injection. The parameter values for the adult are tabulated in Table 1.-1; their basis is discussed below.

(39) *Clearance from the circulation.* After entering the circulation, iron is rapidly bound to plasma transferrin and about half is cleared from the plasma to the euthyroid (red) marrow with a fractional transfer rate of 6 d^{-1} (half-time about 160 min). In the

euthyroid marrow, iron is incorporated into red blood cells and returns to the circulation within a few days ($\lambda = 0.7\ d^{-1}$). At the end of the red blood cells' life (taken to be 120 d in the adult) the iron is released. This process is modelled by a transfer rate from the red blood cell compartment to the "transit" pools of the red marrow of 8.33 $10^{-3}\ d^{-1}$ (mean life of 120 d).

(40) The remaining one-third of plasma iron exchanges about equally with the Liver 1 (transit) compartment and the "rapid turnover" pool of soft tissue. The transfer rates into and out of the latter of 1.4 and 0.91 d^{-1} were based on the MIRD data. The transfer rates into and out of the Liver 1 compartment of 1.5 and 0.01 d^{-1}, respectively, were established to obtain an iron concentration of 0.24 mg g^{-1} in the liver. The transfers of plasma iron in and out of the "transit" red marrow pools assume about 20% of the plasma iron participates in ineffective erythropoiesis.

(41) *Storage iron.* Iron released in the destruction of red blood cells is considered to be stored in the liver (Liver 2), in the marrow ("storage" pool), in the spleen, and in other reticuloendothelial cells ("storage" pool of soft tissue). The fractional transfer rate from the storage sites to the "transit" marrow pools (1.8 $10^{-3}\ d^{-1}$) is taken from the MIRD Report while the rate into these compartments is derived such that the storage corresponds to the following iron levels: spleen 80 mg, liver 50 mg (Liver 2), marrow 300 mg, and soft tissue 340 mg. This assignment is also consistent with the MIRD Report. The "storage" pool of soft tissue also includes, in the adult, about 300 mg of iron incorporated into myoglobin.

(42) *Excretion pathways.* Parameter values for iron losses from the body were chosen for consistency with the summary information of Bothwell *et al.* (1979) where it is suggested that the total losses from the body are normally between about 0.4 and 1 mg d^{-1} for the adult male. The model parameters are based on loss of 0.6 mg d^{-1}. It is assumed that the losses occur from the gastrointestinal tract, the skin, and through urinary excretion in the ratio 6:3:1, respectively. The normal loss of iron by urinary excretion is considered to be less than 0.1 mg d^{-1} and thus a fractional transfer rate of 0.02 d^{-1} from the plasma compartment is assumed. The faecal loss is considered to include a fractional transfer of 1.3 $10^{-4}\ d^{-1}$ from the red blood cell compartment and a contribution from biliary secretion. Biliary secretion is represented in the model by a transfer rate of 4.5 $10^{-4}\ d^{-1}$ from Liver 1 to the small intestine. This rate is based on biliary measurements obtained by T-tube drainage in subjects during and after cholecystectomy, which accounted for approximately 0.24 mg d^{-1} of iron (Green *et al.*, 1968). Biliary iron is assumed to be reabsorbed from the small intestine to blood, with fractional absorption being the same as for directly ingested iron. Losses of iron from the skin were noted to represent about 30% of the total loss. In the model this corresponds to a transfer rate of 5.63 $10^{-4}\ d^{-1}$ from the soft tissue "storage pool".

Parameter values for children

(43) The iron kinetics in children are taken to be similar to the adult. Age-dependent parameters have been established based on estimates of the size of the iron pools in children and a general scaling of adult parameters. The values of the model parameters for children are given in Table 1.-1.

(44) *Iron storage.* The iron storage was estimated at each age. The iron associated with haemoglobin was estimated from the blood mass for different ages (1 g haemoglobin contains 3.4 mg iron). Myoglobin iron was estimated from the muscle mass assumed for each age (0.01 mg iron per g of muscle). Iron levels in the liver and

Table 1.-1. Age-specific transfer rates (d^{-1}) for iron model

	Age					
	3 mo	1 y	5 y	10 y	15 y	Adult
Plasma to R marrow synthesis	1.25E+01	1.80E+01	6.50E+00	6.80E+00	7.10E+00	6.00E+00
Plasma to R marrow transit	1.20E+01	8.50E+00	4.80E+00	3.30E+00	2.25E+00	2.00E+00
Plasma to liver transit	1.50E+00	1.50E+00	1.50E+00	1.50E+00	1.50E+00	1.50E+00
Plasma to S tissue rapid turnover	1.40E+00	1.40E+00	1.40E+00	1.40E+00	1.40E+00	1.40E+00
Plasma to U bladder	1.25E−01	1.00E−01	2.20E−02	2.50E−02	1.85E−02	2.00E−02
RBC to R marrow transit	1.25E−02	1.11E−02	8.33E−03	8.33E−03	8.33E−03	8.33E−03
RBC to ULI	7.50E−04	3.75E−04	1.70E−04	1.80E−04	1.30E−04	1.30E−04
R marrow synthesis to RBC	7.00E−01	7.00E−01	7.00E−01	7.00E−01	7.00E−01	7.00E−01
R marrow transit to plasma	2.00E+00	1.40E+00	7.90E−01	5.40E−01	3.70E−01	3.30E−01
R marrow transit to storage	7.86E−02	8.10E−02	6.32E−02	4.88E−02	3.48E−02	2.47E−02
R marrow transit to liver storage	2.94E−02	2.10E−02	1.19E−02	8.08E−03	5.58E−03	4.95E−03
R marrow transit to spleen	4.69E−02	3.35E−02	1.90E−02	1.29E−02	8.91E−03	7.91E−03
R marrow transit to S tissue storage	7.33E−02	6.10E−02	5.63E−02	4.66E−02	3.48E−02	3.16E−02
Liver transit to plasma	6.00E−03	5.00E−03	1.05E−02	1.28E−02	1.23E−02	1.00E−02
Liver transit to SI content	2.70E−03	1.91E−03	1.08E−03	7.35E−04	5.07E−04	4.50E−04
S tissue rapid turnover to plasma	5.70E−01	4.20E−01	9.70E−01	1.00E+00	1.00E+00	9.10E−01
R marrow storage to R marrow transit	1.10E−02	7.63E−03	4.33E−03	2.94E−03	2.03E−03	1.80E−03
Liver 2 to R marrow transit	1.10E−02	7.63E−03	4.33E−03	2.94E−03	2.03E−03	1.80E−03
Spleen to R marrow transit	1.10E−02	7.63E−03	4.33E−03	2.94E−03	2.03E−03	1.80E−03
S tissue storage to R marrow transit	1.10E−02	7.63E−03	4.33E−03	2.94E−03	2.03E−03	1.80E−03
S tissue storage to excreta	4.29E−03	2.59E−03	9.23E−04	9.23E−04	6.25E−04	5.63E−04
f_1	6.00E−01	2.00E−01	2.00E−01	2.00E−01	2.00E−01	2.00E−01

Parameters are given to sufficient precision for calculational purposes. This may be more precise than the biological data would support.

spleen were estimated from their masses and the assumption that the iron concentrations in these tissues were similar to those in the adult. The partitioning of liver iron between the two liver pools was that of the adult; 90% in Liver 1 and 10% in Liver 2. Based on these assumptions, the following estimates of storage iron were derived:

Table 1.-2. Estimates of storage iron

	Storage iron (mg) at various ages				
Iron pool	3 mo	1 y	5 y	10 y	15 y
Haemoglobin	160	320	710	1300	2300
Myoglobin	14	24	65	130	240
Liver	40	60	130	200	330
Red marrow	20	40	86	160	280

(45) *Iron excretion.* Little is known about the physiological losses of iron in infancy and childhood. Because of the high degree of reutilisation of iron, very little iron is lost from the body on a daily basis, except when bleeding occurs. For example, a 1-y-old infant is estimated to lose about 0.2 mg of iron per day, calculated on the basis of body surface area (Dallman, 1989). Elian *et al.* (1966) estimated that faecal blood losses in children corresponds to about 0.25 mg of iron per d. Garby *et al.* (1964) have estimated losses in children to be 30 μg kg^{-1} which is approximately twice that in the adult. In deriving the age-dependent parameter values, Dallman's suggested scaling of the adult values by surface area. This corresponds to daily iron losses of 0.1, 0.2, 0.3, 0.4, and 0.5 mg for the 3-mo-old infant, and 1, 5, 10, and 15-y-old, respectively. These total losses are distributed between the excretion routes considered in the model in the same manner as the adult.

(46) *Transfer rates.* It is well known that neonatal red blood cells have a shorter life-span than those of the adult. However, limited information is available on the life-span and thus for the 3-mo- and 1-y-old life-spans of 80 and 90 d, respectively have been assumed (Trubowitz and Davis, 1982). For all other ages the life-span is taken to be the

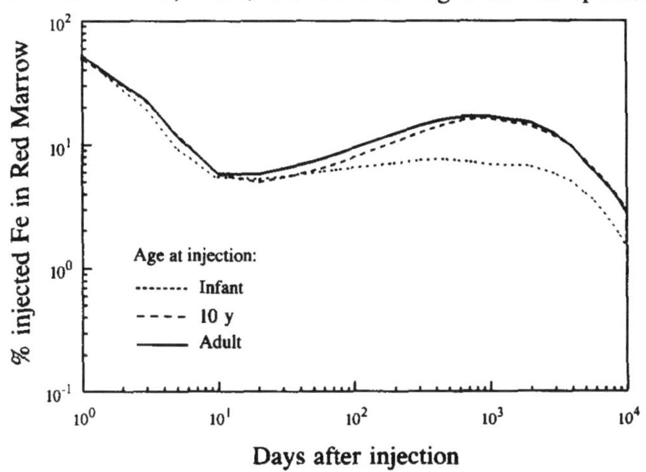

Fig. 1.2. Model predictions of the iron content of the red bone marrow as a function of time after injection into blood for different ages.

same as the adult. Other transfer rates which govern erythropoiesis were assigned values to provide the indicated haemoglobin iron at the various ages. The transfer rates for reutilisation of stored iron were established based on a scaling, by surface area, of the adult values. All other rates were inferred to maintain iron pools of the appropriate size.

(47) Figures 1.2 and 1.3 give estimated iron contents of red bone marrow and liver respectively in infants, 10-y-old children, and adults as a function of time after ingestion based on the parameter values given in Table 1.-1.

Dose coefficients

(47) Dose coefficients derived from the biokinetic data summarised in Table 1.-1 are given in Tables 1.-3 and 1.-4.

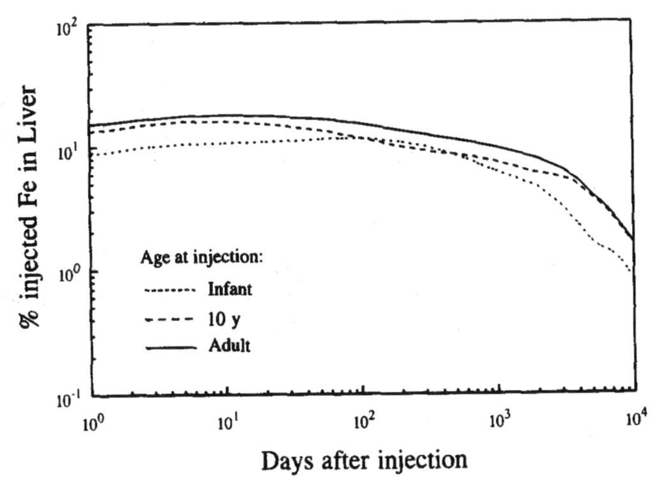

Fig. 1.3. Model predictions of the iron content of the liver as a function of time after injection into blood for different ages.

Table 1.-3.

Ingestion Dose Coefficients: Committed Equivalent and Effective Doses per Unit Intake (Sv/Bq) for Fe-55 (T1/2 - 2.7 y)

Age at intake	3 Months	1 Year	5 Years	10 Years	15 Years	Adult
Adrenals	2.5E-09	7.5E-10	4.6E-10	2.8E-10	1.9E-10	8.6E-11
Bladder Wall	2.5E-09	7.5E-10	4.6E-10	2.8E-10	1.9E-10	8.6E-11
Bone Surfaces	2.1E-08	6.1E-09	4.5E-09	2.9E-09	1.8E-09	6.0E-10
Brain	2.5E-09	7.5E-10	4.6E-10	2.8E-10	1.9E-10	8.6E-11
Breast	2.5E-09	7.5E-10	4.6E-10	2.8E-10	1.9E-10	8.6E-11
GI-Tract						
St Wall	2.6E-09	8.0E-10	4.8E-10	3.0E-10	2.0E-10	9.3E-11
SI Wall	2.6E-09	8.5E-10	5.1E-10	3.1E-10	2.1E-10	1.0E-10
ULI Wall	3.2E-09	1.4E-09	7.8E-10	4.7E-10	3.0E-10	1.8E-10
LLI Wall	4.3E-09	2.6E-09	1.4E-09	8.3E-10	5.0E-10	3.6E-10
Kidneys	2.5E-09	7.5E-10	4.6E-10	2.8E-10	1.9E-10	8.6E-11
Liver	1.6E-08	4.4E-09	3.2E-09	1.9E-09	1.5E-09	7.3E-10
Lungs	2.5E-09	7.5E-10	4.6E-10	2.8E-10	1.9E-10	8.6E-11
Muscle	2.5E-09	7.5E-10	4.6E-10	2.8E-10	1.9E-10	8.6E-11
Ovaries	2.5E-09	7.5E-10	4.6E-10	2.8E-10	1.9E-10	8.6E-11
Pancreas	2.5E-09	7.5E-10	4.6E-10	2.8E-10	1.9E-10	8.6E-11
Red Marrow	2.6E-08	8.0E-09	6.7E-09	4.4E-09	2.9E-09	1.1E-09
Skin	2.5E-09	7.5E-10	4.6E-10	2.8E-10	1.9E-10	8.6E-11
Spleen	4.9E-08	1.5E-08	1.2E-08	8.1E-09	5.9E-09	2.5E-09
Testes	2.5E-09	7.5E-10	4.6E-10	2.8E-10	1.9E-10	8.6E-11
Thymus	2.5E-09	7.5E-10	4.6E-10	2.8E-10	1.9E-10	8.6E-11
Thyroid	2.5E-09	7.5E-10	4.6E-10	2.8E-10	1.9E-10	8.6E-11
Uterus	2.5E-09	7.5E-10	4.6E-10	2.8E-10	1.9E-10	8.6E-11
Remainder	2.6E-08	7.7E-09	6.1E-09	4.2E-09	3.0E-09	1.3E-09
Effective Dose	7.5E-09	2.4E-09	1.7E-09	1.1E-09	7.7E-10	3.3E-10

GI-Tract Gastrointestinal Tract
St Stomach
SI Small Intestine
ULI Upper Large Intestine
LLI Lower Large Intestine

Table 1.-4.

Ingestion Dose Coefficients: Committed Equivalent and Effective Doses per Unit Intake (Sv/Bq) for Fe-59 (T1/2 = 44.529 d)

Age at intake	3 Months	1 Year	5 Years	10 Years	15 Years	Adult
Adrenals	3.1E-08	8.1E-09	5.1E-09	3.5E-09	2.3E-09	1.1E-09
Bladder Wall	2.1E-08	6.8E-09	4.0E-09	2.4E-09	1.6E-09	9.0E-10
Bone Surfaces	7.4E-08	1.6E-08	8.1E-09	4.8E-09	2.9E-09	1.1E-09
Brain	2.0E-08	5.4E-09	2.8E-09	1.8E-09	1.1E-09	4.5E-10
Breast	1.9E-08	4.9E-09	2.7E-09	1.7E-09	1.2E-09	5.0E-10
GI-Tract						
St Wall	2.9E-08	8.7E-09	5.2E-09	3.1E-09	2.1E-09	1.1E-09
SI Wall	3.0E-08	1.3E-08	7.3E-09	4.8E-09	3.1E-09	1.9E-09
ULI Wall	4.0E-08	2.5E-08	1.4E-08	8.6E-09	5.2E-09	4.0E-09
LLI Wall	5.7E-08	5.0E-08	2.6E-08	1.6E-08	9.9E-09	8.4E-09
Kidneys	2.8E-08	7.7E-09	4.8E-09	3.2E-09	2.1E-09	9.8E-10
Liver	7.8E-08	1.8E-08	1.4E-08	9.2E-09	6.2E-09	3.0E-09
Lungs	2.4E-08	6.3E-09	3.6E-09	2.4E-09	1.6E-09	6.9E-10
Muscle	2.2E-08	6.3E-09	3.4E-09	2.3E-09	1.5E-09	7.0E-10
Ovaries	2.8E-08	1.1E-08	6.3E-09	4.3E-09	2.9E-09	1.8E-09
Pancreas	3.4E-08	9.1E-09	5.6E-09	3.6E-09	2.4E-09	1.1E-09
Red Marrow	7.5E-08	1.7E-08	1.0E-08	6.1E-09	3.9E-09	1.6E-09
Skin	1.8E-08	4.8E-09	2.5E-09	1.5E-09	1.0E-09	4.7E-10
Spleen	1.3E-07	2.5E-08	1.5E-08	8.5E-09	5.0E-09	2.0E-09
Testes	2.0E-08	5.7E-09	3.1E-09	2.0E-09	1.3E-09	6.0E-10
Thymus	2.2E-08	5.9E-09	3.2E-09	2.1E-09	1.5E-09	6.1E-10
Thyroid	2.3E-08	6.2E-09	3.3E-09	2.1E-09	1.3E-09	5.5E-10
Uterus	2.6E-08	8.8E-09	4.9E-09	3.3E-09	2.2E-09	1.1E-09
Remainder	7.5E-08	6.4E-09	3.5E-09	2.3E-09	1.5E-09	7.3E-10
Effective Dose	3.9E-08	1.3E-08	7.5E-09	4.7E-09	3.1E-09	1.8E-09

GI-Tract Gastrointestinal Tract
St Stomach
SI Small Intestine
ULI Upper Large Intestine
LLI Lower Large Intestine

References

Bjorn-Rasmussen, E., Hallberg, L., Isaksson, B. and Arvidsson, B. (1974). Food iron absorption in man. Application of the two-pool extrinsic tag method to measure haem and non-haem iron absorption from the whole diet. *J. Clin. Invest.* **53**, 247–255.

Bothwell, T. H., Charlton, R. W., Cook, J. D. and Finch, C. A. (1979). Iron nutrition. In: *Iron Metabolism in Man*, Blackwell Scientific Publications, Oxford.

Cristy, M. and Leggett, R. W. (1986). Determination of metabolic data appropriate for HLW dosimetry. I. Gastrointestinal absorption. *NUREG/CER-3572*, Vol II, pp. 46–51.

Dallman, P. R. (1989). Review of iron metabolism. In: *Dietary Iron: Birth to Two Years* (Filer, L. J. Jr ed.) Raven Press, New York.

Darby, W. J., Hahn, P. F., Kaser, M. M., Steinkamp, R. C., Densen, P. M. and Cook, M. B. (1947). The absorption of radioactive iron by children 7–10 years of age. *J. Nutr.* **33**, 107–119.

Elian, E., Bar-Shani, S., Lierman, A. and Matth, Y. (1966). Intestinal blood loss: A factor in calculations of body iron in late infancy. *J. Pediat.* **69**, 215–219.

Erlandson, M. E., Walden, B., Stern, G., Hilgartner, M. W., Wehman, J. and Smith, C. H. (1962). Studies on congenital hemolytic syndromes. IV. Gastrointestinal absorption of iron. *Blood* **19**, 359–378.

Freiman, H. D., Tauber, S. A. and Tulsky, E. G. (1963). Iron absorption in the healthy aged. *Geriatrics* **18**, 716–720.

Garby, L. and Sjolin, S. (1959). Absorption of labelled iron in infants less than three months old. *Acta Paediat.* **48** (Suppl.), 24–28.

Garby, L., Sjolin, S. and Vuille, J. C. (1964). Studies on erythrokinetics in infancy. IV. The long term behaviour of radio-iron in circulating foetal and adult haemoglobin, and its faecal excretion. *Acta Paediat.* **53**, 33–41.

Gorten, N. K., Hepner, R. and Workman, J. B. (1963). Iron metabolism in premature infants. I. Absorption and utilisation of iron as measured by isotope studies. *J. Pediat.* **63**, 1063–1071.

Green, R., Charlton, R. W., Seftel, H., Bothwell, T., Mayet, F., Adams, B., Finch, C. and Layrisse, M. (1968). Body iron excretion in man. A collaborative study. *Am. J. Med.* **45**, 336–353.

ICRP (1975). *Report of the Task Group on Reference Man.* ICRP Publication 23, Pergamon Press, Oxford.

ICRP (1980). *Limits for Intakes of Radionuclides by Workers.* ICRP Publication 30, Part 2, *Annals of the ICRP* **4**(3/4), Pergamon Press, Oxford.

ICRP (1987). *Radiation Dose to Patients from Radiopharmaceuticals.* ICRP Publication 53, *Annals of the ICRP* **18**(1–4), Pergamon Press, Oxford.

Johnson, J. R. and Dunford, D. W. (1984). Re: Radiation absorbed doses from iron-52, iron-55 and iron-59 used to study ferrokinetics (letter to editor). *J. Nucl. Med.* **25**, 530–531.

Johnson, J. R. and Dunford, D. W. (1985). Comparison of the ICRP and MIRD models for Fe metabolism in man. *Health Phys.* **49**, 211–219.

Layrisse, M. and Martinez-Torres, C. (1971). Iron absorption from food. Iron supplementation of foods. In: *Progress in Hematology*, Vol. VI, pp. 137–160 (Brown, E. B. and Moore, C. V. eds). Grune and Stratton, New York.

Layrisse, M. and Martinez-Torres, C. (1972). Model for measuring dietary absorption of heme iron: Test with a complete meal. *Am. J. Clin. Nutr.* **25**, 401–411.

Martinez-Torres, C. and Layrisse, M. (1973). Nutritional factors in iron deficiency: Food iron absorption. In: *Clinics in Haematology* **2**, pp. 339–352 (Callender, S. T. ed.). W. B. Saunders and Co., London.

Marx, J. J. M. (1979). Normal iron absorption and decreased red cell iron uptake in the aged. *Blood* **53**, 204–211.

NEA/OECD (1988). *Committee on Radiation Protection and Public Health, Expert Group on Gut Transfer Factors.* NEA/OECD, Paris.

Rios, E., Hunter, R. E., Cook, J. D., Smith, N. J. and Finch, C. A. (1975). The absorption of iron as supplements in infant cereal and infant formulas. *Paediatrics* **55**, 694–699.

Robertson, R. R., Price, R. R., Budinger, T. F., Fairbanks, V. F. and Pollycove, M. (1983). Radiation absorbed doses from Iron-52, Iron-55 and Iron-59 used to study ferrokinetics. MIRD Dose Estimate Report No. 11. *J. Nucl. Med.* **24**, 339–348.

Saarinen, U. M. and Siimes, M. A. (1977). Iron absorption from infant milk formula and the optimal level of iron supplementation. *Acta Pediat. Scand.* **66**, 710–722.

Saarinen, U. M., Siimes, M. A. and Dallman, P. R. (1977). Iron absorption in infants: High bioavailability of breast milk iron as indicated by the extrinsic tag method for iron absorption and by the concentration of serum ferritin. *J. Pediat.* **91**, 36–39.

Schulz, J. and Smith, N. J. (1958). A quantitative study of the absorption of food iron in infants and children. *J. Dis. Child.* **95**, 109–119.

Trubowitz, S. and Davis. S. (1982). *The Human Bone Marrow: Anatomy, Physiology, and Pathophysiology.* CRC Press, Inc., Boca Raton, Fl.

2. SELENIUM

Uptake to blood

(48) *Adults.* Selenium (Se) is an essential trace element found as selenocysteine in the enzyme glutathione peroxidase in humans and other mammals and is incorporated into other mammalian proteins. Selenium at high concentrations is also toxic to humans and animals. Several reviews of the behaviour of selenium in the body have been published (Muth *et al.*, 1967; Frost and Lish, 1975; Alexander *et al.*, 1988; Magos and Berg, 1988).

(49) The fractional absorption of selenium in humans and animals has been studied for a variety of conditions including total dietary content of selenium and protein and the chemical form of selenium.

(50) Tracer studies give fractional absorption values between 0.44 and 0.97 (Barbezat *et al.*, 1984a, b). Balance studies, where true absorption has been estimated, gave similar results. Total intake of selenium seems to have little influence on absorption in humans. Values for fractional absorption of about 0.8 have been reported for intakes of 30 and 132 $\mu g\ d^{-1}$ (Stewart *et al.*, 1978; Swanson *et al.*, 1983). Protein content of the diet is known to influence absorption. Greger and Marcus (1981) investigated the effect of protein in the diet and obtained very variable selenium absorption with low protein diets (0.1–0.7) and a higher range for high protein diets (0.6–0.9). The chemical form of selenium plays an important role in absorption. For selenomethionine, which constitutes about 50% of cereal selenium (Thompson *et al.*, 1985), mean fractional absorption values of 0.75 and 0.96 have been reported (Griffiths *et al.*, 1976; Robinson *et al.*, 1978; Bopp *et al.*, 1982). Selenates also appear to be well absorbed but somewhat lower values of about 0.4–0.7 have been reported for selenite (Thompson and Stewart, 1974; Robinson *et al.*, 1978; Janghorbani *et al.*, 1982). The fractional absorption of elemental selenium is low, a value of 0.03 having been reported by Robinson *et al.* (1985) after selenite had been reduced with ascorbic acid.

(51) The fractional absorption of selenium has also been measured in rats, mice, dogs and monkeys. These data indicate that insoluble forms such as selenite are poorly absorbed while selenides, selenates and selenomethionine are well absorbed (Graham *et al.*, 1971; Burk *et al.*, 1972; Thomson and Stewart, 1973; Luckey *et al.*, 1975; Furchner *et al.*, 1975; Nishimura *et al.*, 1991; Archimbaud *et al.*, 1992a). These data indicate f_1 values lying within the range observed in humans.

(52) On the basis of the above data, an f_1 value of 0.8 is adopted here for dietary forms of selenium.

(53) *Children.* Limited data are available on the absorption of selenium in young animals. Nishimura *et al.* (1991) reported that in 6-d-old rats given ^{75}Se as selenite, fractional absorption was much greater than in adults (>0.6). In 14-d-old rats intermediate values were obtained. Since no more relevant data appeared to be available, the same f_1 value as is adopted for adults (0.8) is used here for 1-, 5-, 10-, and 15-y-old children. For 3-mo-old infants an f_1 of 1 is adopted here following the general approach given in Paragraph (8) of the Introduction.

Distribution and retention

(54) *Adults.* The chemical form of selenium injected does not appear to markedly affect its distribution and retention in the body (Cerewenka and Cooper, 1961; Lopez *et al.*, 1969; Thomson and Stewart, 1973).

(55) In rats, the concentration of the stable element resulting from intake in the diet was highest in the kidneys and the testes and lowest in muscle. Overall concentrations in individual tissues varied by about a factor of 10 (Behne *et al.*, 1982). The retention in tissues of ^{75}Se injected in the form of sodium selenite was dependent on the diet. For a diet sufficient in selenium, the concentration in the testes was 1.3 times higher than in the liver. For a selenium-deficient diet the selenium concentration in the testes remained the same but was 14 times that in the liver. This experiment shows that the level of selenium in the male gonads appears to be maintained by regulation mechanisms and that the supply of sufficient amounts of selenium to the testes has priority over that to other tissues. The main part of ^{75}Se taken up by the testes was shown to translocate with the spermatozoa to the epididymis (Behne *et al.*, 1982).

(56) One day after administration to rats of μg amounts of ^{75}Se labelled selenite premixed with sucrose and added to the diet, 16% of the radioactivity was found in urine, 4.8% in GI tract and faeces, 7.0% in liver, 6.3% in kidneys, 3.7% in testes, 1.6% in blood and 61% in carcass. Selenium content in blood, kidneys and the rest of the carcass showed a dependence on previous intake which was not observed for liver. It was concluded that the turnover of selenium in the liver is rapid (Hopkins *et al.*, 1966).

(57) Furchner *et al.* (1975) showed that after intravenous and oral administration of ^{75}Se labelled selenite to mice, rats, dogs and monkeys, the whole body retention could be described by the sum of three exponential components. Nordman (1974) reported that in four patients, followed for 2 mo after intravenous injection of ^{75}Se as the selenite, 40% of the activity was retained in the liver with a biological half-time of 45 d and 14% in the kidneys with a biological half-time of 65 d.

(58) The whole body retention of selenium in man is well described by three exponentials with biological half-times of 0.5–7, 20–70, and 120–330 d (Lathrop *et al.*, 1968, 1972; Falk and Lindhé, 1974; Johnson, 1977; Toohey *et al.*, 1979). Thus Toohey *et al.* (1979) showed, on the basis of radioactive profile scans, that up to 589 d after intravenous injection of ^{75}Se labelled selenomethionine in two humans, no appreciable difference in distribution pattern or retention between the subjects could be observed. Retention could be described by a three-component exponential function with half- times of 0.55 d (13%), 46 d (44%) and 220 d (43%). In a similar experiment Johnson (1977) obtained a long-term retention half-time of 200 d for 32.3% of the injected activity between 76 and 957 d after injection.

(59) Based on the retention parameters given in *ICRP Publications 30* (ICRP, 1980) and *53* (ICRP, 1988), parameters for the tissue distribution adopted here are: liver 25%, kidneys 10%, spleen 1%, pancreas 0.5%, testes 0.1%, ovaries 0.02%, remainder 63.4%. Deposition in testes and ovaries is based on measurements of selenium retention in human tissues (Iyengar *et al.*, 1978; ICRP, 1988). Retention in body tissues following deposition from the transfer compartment is described by the sum of three components with half-times of 3 d (10%), 30 d (40%) and 200 d (50%). The half-time of 200 d is based on the long-term retention of selenomethionine and the assumption that material retained in this compartment is predominantly protein-bound.

(60) Excretion of selenium is mainly in the urine and faeces although it has been reported that small amounts (less than 2% of the administered activity) may be eliminated in the exhaled air (Hopkins *et al.*, 1966; Luckey *et al.*, 1975). The pattern of elimination depends on: the chemical form of selenium and the concentration in the diet; the presence of metals; and the animal species. Selenium excretion is predominantly in the faeces for ruminants (Paulson *et al.*, 1968; Lopez *et al.*, 1968) whereas in monogastric

animals urinary excretion is high, probably due to an effect of sulphur in rumen bacteria (Pope *et al.*, 1979). Intravenously administered selenium is excreted mainly in the urine with some in the faeces in all species (Burk *et al.*, 1972; Shearer and Hadjimarkos, 1973; Luckey *et al.*, 1975).

(61) Hopkins *et al.* (1966) reported a urinary to faecal excretion ratio close to 4 for low dietary concentration of selenium and a ratio of 2 for higher dietary levels. In pregnant rats a urinary to faecal excretion ratio of about 4 was obtained for about 60% of selenium injected as selenite (Shearer and Hadjimarkos, 1973). A similar ratio was obtained by Burk *et al.* (1972) for intravenously injected ^{75}Se after administration of different amounts of stable selenium in the diet. Archimbaud *et al.* (1992a) have given information on temporal changes in urinary to faecal excretion ratio after injection of ^{75}Se as selenite to rats. A urinary to faecal ratio of about 6 was obtained up to 2 d, and 1.5 between 3 and 28 d. No human data appear to be available. A urinary to faecal excretion ratio of 2:1 has been assumed in this report.

(62) *Children.* Limited data appear to be available on the effect of age on selenium distribution and retention. Archimbaud *et al.* (1992b) compared whole body retention in adult and 1-d-old rats after administration of ^{75}Se selenite. Half-times of retention of 1 d (20%) and 30 d (80%) were obtained for neonates compared with 2 d (10%) and 40 d (90%) in adults. This study suggests that retention may be shorter in infants. In the absence of more specific information the biokinetic data used here for adults are also adopted for infants and children.

Dose coefficients

(63) Dose coefficients derived from the biokinetic data summarised in Table 2.-1 are given in Tables 2.-2 and 2.-3.

Table 2.-1. Biokinetic data for selenium

	f_1	Distribution (%)							Biological half-time (d)		
		Liver	Kidneys	Spleen	Pancreas	Testes	Ovaries	Other tissues	Comp.a (10%)	Comp.b (40%)	Comp.c (50%)
3 mo	1	25	10	1	0.5	0.1	0.02	63.4	3	30	200
1 y	0.8	25	10	1	0.5	0.1	0.02	63.4	3	30	200
5 y	0.8	25	10	1	0.5	0.1	0.02	63.4	3	30	200
10 y	0.8	25	10	1	0.5	0.1	0.02	63.4	3	30	200
15 y	0.8	25	10	1	0.5	0.1	0.02	63.4	3	30	200
Adult	0.8	25	10	1	0.5	0.1	0.02	63.4	3	30	200

A urinary to faecal excretion ratio of 2:1 is assumed for selenium that has entered the transfer compartment.

Table 2.-2.

Ingestion Dose Coefficients: Committed Equivalent and Effective Doses per Unit Intake (Sv/Bq) for Se-75 (T1/2 = 119.8 d)

Age at intake	3 Months	1 Year	5 Years	10 Years	15 Years	Adult
Adrenals	2.5E-08	1.6E-08	9.9E-09	7.0E-09	4.8E-09	4.0E-09
Bladder Wall	1.1E-08	6.8E-09	4.1E-09	2.7E-09	1.9E-09	1.5E-09
Bone Surfaces	1.6E-08	1.0E-08	5.8E-09	4.0E-09	2.7E-09	2.4E-09
Brain	8.4E-09	5.3E-09	3.1E-09	1.9E-09	1.2E-09	1.0E-09
Breast	8.7E-09	5.4E-09	3.1E-09	2.0E-09	1.3E-09	1.1E-09
GI-Tract						
St Wall	1.7E-08	1.1E-08	6.6E-09	4.4E-09	2.8E-09	2.3E-09
SI Wall	1.7E-08	1.1E-08	6.6E-09	4.3E-09	2.7E-09	2.3E-09
ULI Wall	1.8E-08	1.3E-08	7.6E-09	4.7E-09	3.1E-09	2.5E-09
LLI Wall	1.4E-08	1.1E-08	6.3E-09	4.1E-09	2.6E-09	2.2E-09
Kidneys	8.3E-08	5.1E-08	3.2E-08	2.2E-08	1.6E-08	1.4E-08
Liver	6.0E-08	3.9E-08	2.3E-08	1.7E-08	1.2E-08	1.0E-08
Lungs	1.4E-08	8.7E-09	5.1E-09	3.5E-09	2.5E-09	2.0E-09
Muscle	1.1E-08	7.1E-09	4.0E-09	2.7E-09	1.8E-09	1.5E-09
Ovaries	1.9E-08	1.2E-08	6.8E-09	4.3E-09	2.4E-09	2.1E-09
Pancreas	4.0E-08	2.4E-08	1.4E-08	1.0E-08	6.1E-09	4.8E-09
Red Marrow	1.0E-08	6.5E-09	4.1E-09	2.9E-09	2.0E-09	1.8E-09
Skin	7.3E-09	4.5E-09	2.5E-09	1.6E-09	1.1E-09	9.5E-10
Spleen	3.2E-08	2.0E-08	1.2E-08	8.2E-09	5.5E-09	4.3E-09
Testes	2.6E-08	1.9E-08	1.4E-08	1.1E-08	2.6E-09	1.7E-09
Thymus	1.0E-08	6.6E-09	3.8E-09	2.5E-09	1.7E-09	1.4E-09
Thyroid	1.1E-08	6.9E-09	4.0E-09	2.5E-09	1.6E-09	1.3E-09
Uterus	1.3E-08	8.7E-09	5.1E-09	3.3E-09	2.2E-09	1.8E-09
Remainder	4.7E-08	2.9E-08	1.8E-08	1.3E-08	9.2E-09	7.6E-09
Effective Dose	2.0E-08	1.3E-08	8.4E-09	6.1E-09	3.2E-09	2.6E-09

GI-Tract Gastrointestinal Tract
St Stomach
SI Small Intestine
ULI Upper Large Intestine
LLI Lower Large Intestine

Table 2.-3.

Ingestion Dose Coefficients: Committed Equivalent and Effective Doses per Unit Intake (Sv/Bq) for Se-79 (T1/2 = 65000 y)

Age at intake	3 Months	1 Year	5 Years	10 Years	15 Years	Adult
Adrenals	1.0E-08	6.1E-09	3.2E-09	1.9E-09	1.2E-09	9.7E-10
Bladder Wall	1.0E-08	6.3E-09	3.5E-09	2.1E-09	1.3E-09	1.1E-09
Bone Surfaces	1.0E-08	6.1E-09	3.2E-09	1.9E-09	1.2E-09	9.7E-10
Brain	1.0E-08	6.1E-09	3.2E-09	1.9E-09	1.2E-09	9.7E-10
Breast	1.0E-08	6.1E-09	3.2E-09	1.9E-09	1.2E-09	9.7E-10
GI-Tract						
St Wall	1.1E-08	6.5E-09	3.5E-09	2.0E-09	1.2E-09	1.0E-09
SI Wall	1.0E-08	6.3E-09	3.4E-09	2.0E-09	1.2E-09	1.0E-09
ULI Wall	1.3E-08	9.4E-09	4.9E-09	2.9E-09	1.7E-09	1.4E-09
LLI Wall	1.9E-08	1.5E-08	8.1E-09	4.8E-09	2.8E-09	2.3E-09
Kidneys	2.3E-07	1.4E-07	8.2E-08	5.5E-08	3.9E-08	3.2E-08
Liver	1.2E-07	7.4E-08	4.1E-08	2.7E-08	1.7E-08	1.4E-08
Lungs	1.0E-08	6.1E-09	3.2E-09	1.9E-09	1.2E-09	9.7E-10
Muscle	1.0E-08	6.1E-09	3.2E-09	1.9E-09	1.2E-09	9.7E-10
Ovaries	3.7E-08	2.4E-08	1.1E-08	5.8E-09	1.8E-09	1.8E-09
Pancreas	7.5E-08	4.1E-08	2.1E-08	1.5E-08	7.3E-09	5.0E-09
Red Marrow	1.0E-08	6.1E-09	3.2E-09	1.9E-09	1.2E-09	9.7E-10
Skin	1.0E-08	6.1E-09	3.2E-09	1.9E-09	1.2E-09	9.7E-10
Spleen	5.6E-08	3.4E-08	2.0E-08	1.2E-08	7.8E-09	5.6E-09
Testes	1.0E-07	7.7E-08	6.0E-08	4.6E-08	5.9E-09	2.9E-09
Thymus	1.0E-08	6.1E-09	3.2E-09	1.9E-09	1.2E-09	9.7E-10
Thyroid	1.0E-08	6.1E-09	3.2E-09	1.9E-09	1.2E-09	9.7E-10
Uterus	1.0E-08	6.1E-09	3.2E-09	1.9E-09	1.2E-09	9.7E-10
Remainder	1.2E-07	7.3E-08	4.3E-08	2.9E-08	2.0E-08	1.7E-08
Effective Dose	4.1E-08	2.8E-08	1.9E-08	1.4E-08	4.0E-09	2.9E-09

GI-Tract Gastrointestinal Tract
St Stomach
SI Small Intestine
ULI Upper Large Intestine
LLI Lower Large Intestine

References

Alexander, J., Högberg, J., Thomassen, Y. and Aaseth, J. (1988). Selenium. In: *Handbook on Toxicity of Inorganic Compounds*, pp. 581–594 (Seiler, H. G., Sigel, H. and Sigel, A. eds). Marcel Dekker, New York.

Archimbaud, Y., Grillon, G., Poncy, J. L. and Masse, R. (1992a). Toxicocinetique du selenium 75 et du technetium 95m chez le rat en fonction de l'age et au cours de la gestation et de la lactation. *Rapport CEA-R-5599*, Gif-sur-Yvette, France.

Archimbaud, Y., Grillon, G., Poncy, J. L. and Masse, R. (1992b). ^{75}Se transfer via placenta and milk, distribution and retention in fetal, young and adult rat. *Radiat. Prot. Dosim.* **41**, 147–151.

Barbezat, G. O., Casey, C. E., Reasbeck, P. G., Robinson, M. F. and Thomson, C. D. (1984a). Selenium. In: *Absorption and Malabsorption of Mineral Nutrients*, pp. 231–258. Alan R. Liss, Inc.

Barbezat, G. O., Casey, C. E., Reasbeck, P. G. and Robinson, M. F. (1984b). Selenium. *Curr. Topics Nutr. Dis.* **12**, 231–258.

Behne, D., Höfer, T., von Berswordt-Wallrabe, R. and Elger, W. (1982). Selenium in the testis of the rat: Studies on its regulation and its importance for the organism. *J. Nutr.* **112**, 1682–1687.

Bopp, B. A., Sonders, R. C. and Kesteron, J. W. (1982). Metabolic fate of selected selenium compounds in laboratory animals and man. *J. Drug Metab. Rev.* **13**, 271–318.

Burk, R. F., Brown, D. G., Seely, R. J. and Scaieff, C. C. III (1972). Influence of dietary and injected selenium on whole-body retention, route of excretion and tissue retention of $^{75}SeO_3^{2-}$ in the rat. *J. Nutr.* **102**, 1049–1056.

Cerwenka, E. A. Jr and Cooper, W. C. (1961). Toxicology of selenium and tellurium and their compounds. *Arch. Environ. Health* **3**, 189–200.

Falk, R. and Lindhé, J. C. (1974). Radiation dose received by humans from intravenously administered sodium selenite marked with selenium-75. *SSI-1974-011, also LA-tr-75-6.*

Frost, D. F. and Lish, P. M. (1975). Selenium in biology. *Ann. Rev. Pharmacol.* **15**, 259–284.

Furchner, J. E., London, J. E. and Wilson, J. S. (1975). Comparative metabolism of radionuclides in mammals—IX. Retention of ^{75}Se in the mouse, rat, monkey and dog. *Health Phys.* **29**, 641–648.

Graham, L. A., Veatch, R. L. and Kaplan, E. (1971). Distribution of ^{75}Se-selenomethionine as influenced by the route of administration. *J. Nucl. Med.* **12**, 566–569.

Greger, J. and Marcus, R. (1981). Effect of dietary protein, phosphorus and sulphur amino acids on selenium metabolism of adult males. *Ann. Nutr. Metab.* **25**, 97–108.

Griffiths, N., Stewart, R. and Robinson, M. (1976). The metabolism of selenomethionine in four women. *Br. J. Nutr.* **35**, 373–383.

Hopkins, L. L. Jr, Pope, A. L. and Baumann, C. A. (1966). Distribution of microgram quantities of selenium in the tissues of the rat, and effect of previous selenium intake. *J. Nutr.* **88**, 61–65.

ICRP (1980). *Limits for Intakes of Radionuclides by Workers.* ICRP Publication 30, Part 2. *Annals of the ICRP* **4**(3/4), Pergamon Press, Oxford.

ICRP (1988). *Radiation Dose to Patients from Radiopharmaceuticals.* ICRP Publication 53. *Annals of the ICRP* **18**(1–4), Pergamon Press, Oxford.

Iyengar, G. V., Kollmer, W. E. and Bowen, H. J. M. (1978). *The Elemental Composition of Human Tissues and Body Fluids.* Verlag Chemie, Weinheim.

Janghorbani, M., Christensen, M., Nahapetian, A. and Young, V. (1982). Selenium metabolism in healthy adults; quantitative aspects using the stable isotope SeO_3. *Am. J. Clin. Nutr.* **35**, 647–654.

Johnson, J. R. (1977). Whole body retention following an intravenous injection of ^{75}Se as selenomethionine. *Health Phys.* **33**, 250–251.

Lathrop, K., Harper, P. V. and Malkinson, F. D. (1968). Human total-body retention and excretory routes of ^{75}Se from selenomethionine. *Strahlentherapie* **56**, 436–443.

Lathrop, K. A., Johnston, R. E., Blau, M. and Rothschild, E. O. (1972). Radiation dose to humans from ^{75}Se-L-selenomethionine. *J. Nucl. Med.* **13** (Suppl. 6).

Lopez, P. L., Preston, R. L. and Pfander, W. H. (1968). Whole body retention, tissue distribution and excretion of Selenium-75 after oral and intravenous administration in lambs fed varying selenium intakes. *J. Nutr.* **97**, 123–132.

Luckey, T. D., Venugopal, B. and Hutcheson, D. (1975). *Heavy Metal Toxicity, Safety and Hormonology.* Georg Thieme Publishers, Stuttgart.

Magos, L. and Berg, G. G. (1988). Selenium. In: *Biological Monitoring of Toxic Metals* (Clarkson, T. W., Fuberg, L., Nordberg, G. F. and Sager, P. R. eds). Plenum Press, New York.

Muth, O. H., Oldfield, J. E. and Weswig, P. H. (eds) (1967). *Selenium in Biomedicine.* Westport CT, AVI.

Nishimura, Y., Inaba, J., Matusaka, N. and Danbara, H. (1991). Biokinetics of selenium in rats of various ages. *Biomed. Res. Trace Elements* **2**, 11–19.

Nordman, E. (1974). Measurements of ^{75}Se in the human body. *Acta Radiol. (Suppl.)* **340**, 59–62.

Paulson, G. D., Baumann, C. A. and Pope, A. L. (1968). Metabolism of ^{75}Se-selenite, ^{75}Se-selenate, ^{75}Se-selenomethionine and ^{35}S-sulfate by rumen microorganisms *in vitro. J. Anim. Sci.* **27**, 497–504.

Pope, A. L., Moir, R. J., Somers, M., Underwood, E. J. and White, C. L. (1979). The effect of sulphur on ^{75}Se absorption and retention in sheep. *J. Nutr.* **109**, 1448–1455.

Robinson, J. R., Robinson, M. F., Levander, O. A. and Thomson, C. D. (1985). Urinary excretion of selenium by New Zealand and North American human subjects on differing intakes. *Am. J. Clin. Nutr.* **41**, 1023–1031.

Robinson, M., Rea, H., Friend, G., Stewart, R., Snow, P. and Thomson, C. (1978). On supplementing the selenium intake of New Zealanders. *Br. J. Nutr.* **39**, 589–600.

Shearer, T. R. and Hadjimarkos, D. M. (1973). Comparative distribution of ^{75}Se in the hard and soft tissues of mother rats and their pups. *J. Nutr.* **103**, 553–559.

Stewart, R., Griffiths, N., Thomson, C. and Robinson, M. (1978). Quantitative selenium metabolism in normal New Zealand women. *Br. J. Nutr.* **40**, 45–54.

Swanson, C., Reamer, D., Veillon, C., King, J. and Levander, O. (1983). Quantitative and qualitative aspects of selenium utilization in pregnant and nonpregnant women — an application of stable isotope methodology. *Am. J. Clin. Nutr.* **38**, 169–180.

Thomson, C. D. and Stewart, R. D. H. (1973). Metabolic studies of [^{75}Se] selenomethionne and [^{75}Se] selenite in the rat. *J. Nutr.* **30**, 139–147.

Thomson, C. D. and Stewart, R. D. H. (1974). The metabolism of ^{75}Se selenite in young women. *Br. J. Nutr.* **32**, 303–323.

Thomson, C. D., Ong, L. K. and Robinson, M. F. (1985). Effects of supplementation with high-selenium wheat bread on selenium, glutathione peroxidase and related enzymes in blood components of New Zealand residents. *Am. J. Clin. Nutr.* **41**, 1015–1022.

Toohey, R. E., Essling, M. A. and Huff, D. R. (1979). Retention and gross distribution of ^{75}Se following intravenous injection of ^{75}Se-selenomethionine. *Health Phys.* **37**, 395–397.

3. ANTIMONY

Uptake to blood

(64) *Adults.* No experimental studies of antimony (Sb) absorption in humans appear to have been carried out. Bioassay measurements following an accidental exposure to antimony-containing dust suggested that absorption was less than 5% (Coughtrey and Thorne, 1983; Bailly *et al.*, 1991). Variable results have been reported on the absorption of antimony from the gastrointestinal tract in experimental animals. Waitz *et al.* (1965) reported that in mice, fractional absorption of antimony from antimonyl potassium tartrate (tartar emetic) was about 0.2. Moskalev (1964), however, found fractional absorption of only about 0.05 from this compound in rats and Felicetti *et al.* (1974a) reported that the fractional absorption of trivalent and pentavalent ^{124}Sb–tartrate complexes in hamsters was less than 1%. Absorption of other forms of antimony is apparently much less than 10% (Rose and Jacobs, 1969; Thomas *et al.*, 1973). Based on these studies, the International Commission on Radiological Protection (ICRP, 1981) recommended f_1 values of 0.1 for tartar emetic and 0.01 for all other compounds of antimony.

(65) Van Bruwaene *et al.* (1982) orally administered $^{124}SbCl_3$ to lactating dairy cows. They reported that the excretion of antimony into urine after oral administration was about 1%, whereas about 51% was recovered from urine after intravenous injection; about 82% of orally administered antimony and about 2.4% of intravenously administered antimony was recovered in faeces. These results are difficult to interpret but suggest an f_1 value in the range between about 0.02 and 0.2. Chertok and Lake (1970) reported that for dogs fed antimony in debris from a sub-surface nuclear test, fractional absorption was at least 0.04. Moreover, much of the antimony in the ingested material may have been unavailable for absorption. Inaba *et al.* (1984a) reported fractional absorption in adult rats as 0.5 of ^{125}Sb biologically incorporated into blood cells. In a further study in which ^{125}Sb was mixed with blood, fractional absorption was only about 0.01; however, hydrolysis may have occurred which would have reduced uptake.

(66) Coughtrey and Thorne (1983) have suggested an upper limit for the f_1 value of about 0.1, based on estimates of the daily dietary intake and the body content of stable antimony.

(67) These data indicate there may be considerable differences in absorption for the range of chemical forms of antimony encountered in the environment. Because ingestion of radioantimony by the general public is likely to occur principally in food or drink, an NEA/OECD Expert Group (1988) assumed the higher value of 0.1 adopted in *ICRP Publication 30* (ICRP, 1981) for the fractional absorption of antimony. This f_1 value is also adopted here.

(68) *Children.* Few data seem to be available on the absorption of antimony in young animals. Inaba *et al.* (1984b) administered $^{125}SbCl_3$ to 5-d-old suckling rats and to adult rats, and compared their whole body retention of ^{125}Sb. Retention of antimony on the fifth day after administration was about 40% for suckling animals and 0.2% for adults. For 15-d-old suckling animals, retention on the fifth day was 20% and for 25-d-old weanlings it was about the same as for adults (Inaba *et al.*, 1984a).

(69) Because there appears to be no further information available concerning intestinal absorption of antimony in human infants and neonates, the general approach of

the NEA expert group (NEA/OECD, 1988) is adopted here (see Paragraph (8) of the Introduction) to give an f_1 value for the 3-mo-old infant of 0.2. For children of 1 y and older, the f_1 value for the adult of 0.1 is used here.

Distribution and retention

(70) *Adults.* Iyengar *et al.* (1978) reviewed data on the concentration of stable antimony in human tissues. Concentrations were highest in hair, bone and teeth.

(71) The tissue distribution of antimony appears to be species-dependent. Trivalent antimony concentrates in the liver of many animal species (Moskalev, 1964; Waitz *et al.*, 1965; Felicetti *et al.*, 1974b; Berman *et al.*, 1988), and in the thyroid and parathyroid of dogs (Felicetti *et al.*, 1974b). In rats, red blood cells accumulate a large percentage of total body antimony (Djurić *et al.*, 1962; Tarrant *et al.*, 1971; Inaba *et al.*, 1984a). Accumulation of antimony in the skin and skeleton have been reported in mice (Thomas *et al.*, 1973; Tarrant *et al.*, 1971) and hamsters (Berman *et al.*, 1988). In lactating dairy cows it is reported that the spleen accumulated ^{124}Sb after oral administration and little was secreted into milk (Van Bruwaene *et al.*, 1982).

(72) Coughtrey and Thorne (1983) developed a model for the whole body retention of antimony, which incorporated into the *ICRP Publication 30* (1981) model an additional component of retention with a longer biological half-time to account for the content of stable antimony in man (Iyengar *et al.*, 1978). It also allowed for the pattern of retention of antimony in a worker who had been accidentally exposed (Rose and Jacobs, 1969).

(73) The model of Coughtrey and Thorne is used as a basis for the model adopted in this report. The parameter values have been rounded because of the considerable uncertainties caused by the wide differences of published data (Moskalev, 1964; Thomas *et al.*, 1973; van Bruwaene *et al.*, 1982; Bailly *et al.*, 1991). It is therefore assumed that of antimony leaving the transfer compartment, 0.2 goes directly to excretion, 0.4 is translocated to mineral bone, 0.05 is translocated to the liver, and the remaining fraction (0.35) is uniformly distributed throughout all other organs and tissues of the body. Of antimony translocated to any organ or tissue, fractions of 0.85, 0.1, and 0.05 are assumed to be retained with biological half-times of 5, 100, and 5000 d, respectively.

(74) On the basis of human data on the excretion of stable antimony (ICRP, 1975) and rat data (Moskalev, 1964) a urinary to faecal excretion ratio of 4:1 is assumed here.

(75) *Children.* No information appears to be available on the age-dependent tissue distribution of antimony in humans. In rats, whole body retention of ^{125}Sb administered as $SbCl_3$ appears to be independent of age (Inaba *et al.*, 1984b). In the absence of further information the retention parameters adopted for adults are also used here for infants and children.

Classification of isotopes for bone dosimetry

(76) In *ICRP Publication 30,* it was assumed that isotopes of antimony are uniformly retained on bone surfaces. Because no data have been found regarding the distribution of isotopes of antimony in the skeleton, this assumption is also adopted here.

Dose coefficients

(77) Dose coefficients derived from the biokinetic data summarised in Table 3.-1 are given in Tables 3.-2 to 3.-4.

Table 3.-1. Biokinetic data for antimony

		Distribution (%)				Biological half-time (*d*)		
	f_1	Skeleton	Liver	Other tissues	Prompt excretion	Comp. A (85%)	Comp. B (10%)	Comp. C (5%)
3 mo	0.2	40	5	35	20	5	100	5000
1 y	0.1	40	5	35	20	5	100	5000
5 y	0.1	40	5	35	20	5	100	5000
10 y	0.1	40	5	35	20	5	100	5000
15 y	0.1	40	5	35	20	5	100	5000
Adult	0.1	40	5	35	20	5	100	5000

A urinary to faecal excretion ratio of 4:1 is assumed for antimony that has entered the transfer compartment.

Table 3.-2.

Ingestion Dose Coefficients: Committed Equivalent and Effective Doses per Unit Intake (Sv/Bq) for Sb-124 (T1/2 = 60.20 d)

Age at intake	3 Months	1 Year	5 Years	10 Years	15 Years	Adult
Adrenals	4.1E-09	1.9E-09	1.1E-09	7.0E-10	4.8E-10	3.9E-10
Bladder Wall	6.2E-09	3.8E-09	2.3E-09	1.6E-09	1.0E-09	8.6E-10
Bone Surfaces	3.0E-08	1.3E-08	8.5E-09	4.8E-09	2.8E-09	2.7E-09
Brain	2.6E-09	9.6E-10	5.2E-10	3.3E-10	2.2E-10	1.8E-10
Breast	2.4E-09	9.9E-10	5.2E-10	3.3E-10	2.0E-10	1.7E-10
GI-Tract						
St Wall	1.1E-08	6.4E-09	3.3E-09	2.0E-09	1.4E-09	1.1E-09
SI Wall	2.1E-08	1.5E-08	8.3E-09	5.3E-09	3.3E-09	2.6E-09
ULI Wall	8.0E-08	5.8E-08	3.0E-08	1.8E-08	1.1E-08	8.6E-09
LLI Wall	2.1E-07	1.5E-07	7.6E-08	4.6E-08	2.7E-08	2.2E-08
Kidneys	4.1E-09	2.1E-09	1.3E-09	8.4E-10	5.5E-10	4.6E-10
Liver	7.8E-09	3.5E-09	1.9E-09	1.2E-09	7.9E-10	6.4E-10
Lungs	3.0E-09	1.3E-09	6.7E-10	4.3E-10	2.8E-10	2.3E-10
Muscle	3.7E-09	1.9E-09	1.1E-09	7.0E-10	4.6E-10	3.8E-10
Ovaries	1.2E-08	9.1E-09	5.2E-09	3.6E-09	2.5E-09	1.9E-09
Pancreas	4.4E-09	2.3E-09	1.3E-09	8.4E-10	5.4E-10	4.4E-10
Red Marrow	2.1E-08	7.3E-09	3.9E-09	2.4E-09	1.6E-09	1.2E-09
Skin	2.7E-09	1.2E-09	6.4E-10	4.0E-10	2.6E-10	2.2E-10
Spleen	3.8E-09	1.9E-09	1.1E-09	6.7E-10	4.3E-10	3.5E-10
Testes	3.6E-09	1.9E-09	1.1E-09	6.7E-10	4.1E-10	3.2E-10
Thymus	2.6E-09	1.0E-09	5.6E-10	3.5E-10	2.4E-10	2.0E-10
Thyroid	2.6E-09	9.9E-10	5.4E-10	3.4E-10	2.2E-10	1.9E-10
Uterus	7.4E-09	5.1E-09	2.8E-09	1.9E-09	1.2E-09	9.3E-10
Remainder	3.9E-09	2.0E-09	1.2E-09	7.6E-10	5.1E-10	4.2E-10
Effective Dose	2.5E-08	1.6E-08	8.4E-09	5.2E-09	3.2E-09	2.5E-09

GI-Tract Gastrointestinal Tract
St Stomach
SI Small Intestine
ULI Upper Large Intestine
LLI Lower Large Intestine

Table 3.-3.

Ingestion Dose Coefficients: Committed Equivalent and Effective Doses per Unit Intake (Sv/Bq) for Sb-125 (T1/2 = 2.77 y)

Age at intake	3 Months	1 Year	5 Years	10 Years	15 Years	Adult
Adrenals	3.5E-09	1.5E-09	9.7E-10	6.6E-10	4.8E-10	4.3E-10
Bladder Wall	3.3E-09	1.7E-09	1.1E-09	7.6E-10	5.0E-10	4.4E-10
Bone Surfaces	7.4E-08	3.3E-08	2.3E-08	1.3E-08	9.2E-09	9.1E-09
Brain	2.5E-09	1.0E-09	6.4E-10	4.1E-10	2.9E-10	2.6E-10
Breast	2.0E-09	8.5E-10	5.1E-10	3.3E-10	2.3E-10	2.1E-10
GI-Tract						
St Wall	4.7E-09	2.4E-09	1.4E-09	8.5E-10	5.9E-10	5.0E-10
SI Wall	7.7E-09	5.0E-09	2.9E-09	1.9E-09	1.2E-09	9.8E-10
ULI Wall	2.3E-08	1.6E-08	8.6E-09	5.3E-09	3.1E-09	2.5E-09
LLI Wall	5.8E-08	4.1E-08	2.1E-08	1.3E-08	7.6E-09	6.2E-09
Kidneys	3.1E-09	1.4E-09	9.2E-10	6.1E-10	4.3E-10	3.8E-10
Liver	7.0E-09	3.0E-09	1.9E-09	1.3E-09	9.0E-10	7.9E-10
Lungs	2.7E-09	1.1E-09	6.9E-10	4.5E-10	3.3E-10	2.9E-10
Muscle	2.8E-09	1.3E-09	7.7E-10	5.0E-10	3.5E-10	3.1E-10
Ovaries	5.6E-09	3.6E-09	2.2E-09	1.5E-09	1.0E-09	8.0E-10
Pancreas	3.3E-09	1.5E-09	9.4E-10	6.2E-10	4.4E-10	3.9E-10
Red Marrow	2.0E-08	7.7E-09	4.4E-09	2.6E-09	1.8E-09	1.5E-09
Skin	2.1E-09	9.1E-10	5.4E-10	3.4E-10	2.4E-10	2.1E-10
Spleen	2.8E-09	1.3E-09	7.8E-10	5.1E-10	3.5E-10	3.1E-10
Testes	2.5E-09	1.2E-09	6.8E-10	4.4E-10	2.9E-10	2.6E-10
Thymus	2.3E-09	9.8E-10	6.0E-10	3.9E-10	2.8E-10	2.5E-10
Thyroid	2.4E-09	1.0E-09	6.2E-10	4.0E-10	2.9E-10	2.6E-10
Uterus	3.9E-09	2.2E-09	1.4E-09	8.8E-10	5.9E-10	5.0E-10
Remainder	2.8E-09	1.3E-09	8.0E-10	5.2E-10	3.7E-10	3.2E-10
Effective Dose	1.1E-08	6.1E-09	3.4E-09	2.1E-09	1.4E-09	1.1E-09

GI-Tract Gastrointestinal Tract
St Stomach
SI Small Intestine
ULI Upper Large Intestine
LLI Lower Large Intestine

Table 3.-4.

Ingestion Dose Coefficients: Committed Equivalent and Effective Doses per Unit Intake (Sv/Bq) for Sb-126 (T1/2 = 12.4 d)

Age at intake	3 Months	1 Year	5 Years	10 Years	15 Years	Adult
Adrenals	3.5E-09	1.9E-09	1.1E-09	6.7E-10	4.5E-10	3.6E-10
Bladder Wall	7.1E-09	5.0E-09	3.0E-09	2.2E-09	1.3E-09	1.1E-09
Bone Surfaces	1.1E-08	5.0E-09	3.2E-09	1.9E-09	1.1E-09	1.1E-09
Brain	1.6E-09	6.0E-10	3.3E-10	2.1E-10	1.4E-10	1.2E-10
Breast	1.7E-09	7.8E-10	4.1E-10	2.5E-10	1.5E-10	1.2E-10
GI-Tract						
St Wall	1.1E-08	6.6E-09	3.6E-09	2.3E-09	1.5E-09	1.2E-09
SI Wall	2.2E-08	1.7E-08	9.6E-09	6.3E-09	4.0E-09	3.2E-09
ULI Wall	6.7E-08	4.9E-08	2.6E-08	1.7E-08	9.9E-09	8.0E-09
LLI Wall	1.6E-07	1.2E-07	6.1E-08	3.7E-08	2.2E-08	1.8E-08
Kidneys	3.8E-09	2.3E-09	1.4E-09	9.5E-10	6.2E-10	5.2E-10
Liver	5.5E-09	3.0E-09	1.7E-09	1.0E-09	6.3E-10	5.1E-10
Lungs	2.2E-09	1.1E-09	5.4E-10	3.5E-10	2.3E-10	1.8E-10
Muscle	3.5E-09	2.1E-09	1.2E-09	7.9E-10	5.2E-10	4.3E-10
Ovaries	1.7E-08	1.4E-08	8.0E-09	5.5E-09	3.8E-09	2.8E-09
Pancreas	4.4E-09	2.6E-09	1.5E-09	9.7E-10	6.1E-10	4.9E-10
Red Marrow	9.5E-09	3.7E-09	2.3E-09	1.6E-09	1.1E-09	9.2E-10
Skin	2.1E-09	1.1E-09	5.8E-10	3.6E-10	2.3E-10	1.9E-10
Spleen	3.6E-09	2.1E-09	1.2E-09	7.6E-10	4.7E-10	3.8E-10
Testes	3.7E-09	2.4E-09	1.3E-09	7.8E-10	4.4E-10	3.6E-10
Thymus	1.8E-09	7.3E-10	4.1E-10	2.5E-10	1.7E-10	1.4E-10
Thyroid	1.7E-09	6.8E-10	3.7E-10	2.3E-10	1.5E-10	1.2E-10
Uterus	9.3E-09	7.1E-09	4.2E-09	2.7E-09	1.7E-09	1.3E-09
Remainder	3.4E-09	2.1E-09	1.2E-09	8.5E-10	5.8E-10	4.7E-10
Effective Dose	2.0E-08	1.4E-08	7.7E-09	4.9E-09	3.1E-09	2.5E-09

GI-Tract Gastrointestinal Tract
St Stomach
SI Small Intestine
ULI Upper Large Intestine
LLI Lower Large Intestine

References

Bailly, R., Lauwerys, R., Bucher, J. P., Mahieu, P. and Konings, J. (1991). Experimental and human studies on antimony metabolism: Their relevance for the biological monitoring of workers exposed to inorganic antimony. *Br. J. Indust. Med.* **48**, 93–97.

Berman, J. D., Gallaler, J. F. and Gallaler, J. V. (1988). Pharmacokinetics of pentavalent antimony (Pentostan) in hamsters. *Am. J. Trop. Med. Hyg.* **39**, 41–45.

Chertok, R. J. and Lake, S. (1970). Availability in the dog of radionuclides in nuclear debris from the Plowshare excavation Cabriolet. *Health Phys.* **19**, 405–409.

Coughtrey, P. J. and Thorne, M. C. (1983). *Radionuclide Distribution and Transport in Terrestrial and Aquatic Ecosystems*, Vol. 3 (Balkema, A. A. ed.), Rotterdam.

Djurić, D., Thomas, R. G. and Lie, R. (1962). The distribution and excretion of trivalent antimony in the rat following inhalation. *Int. Arch. Gewerbepathol. Gewerbehyg.* **19**, 529–545.

Felicetti, S. A., Thomas, R. G. and McClellan, R. O. (1974a). Metabolism of two valence states of inhaled antimony in hamsters. *Am. Ind. Hyg. Assoc. J.* **35**, 292–300.

Felicetti, S. A., Thomas, R. G. and McClellan, R. O. (1974b). Retention of inhaled antimony-124 in the beagle dog as a function of temperature of aerosol formation. *Health Phys.* **26**, 525–531.

ICRP (1981). *Limits for Intakes of Radionuclides by Workers.* ICRP Publication 30, Part 3. *Annals of the ICRP* **6**(2/3), Pergamon Press, Oxford.

Inaba, J., Nishimura, Y. and Ichikawa, R. (1984a). Studies on the metabolism of antimony-125 in the rat. *NIRS-N-49 (National Institute of Radiological Sciences, Chiba, Japan)*, 81–83.

Inaba, J., Nishimura, Y. and Ichikawa, R. (1984b). Effect of age on the metabolism of some important radionuclides in the rat. In: *Radiation Risk Protection.* Proc. 6th Int. Congr. of IRPA, Vol. 1, pp. 481–484 (Kaul, A., Neider, R., Pensko, J., Stieve, F. E. and Brunner, H. eds.) Fachverband für Strahlenschutz e. V., Berlin.

Iyengar, G. V., Kellner, W. E. and Bauer, H. J. M. (1978). *The Elemental Composition of Human Tissues and Body Fluids.* Verlag Chemie, Weinheim.

Moskalev, Y. I. (1964). Experiments dealing with distribution of antimony-124 and tellurium-127. *AEC-tr-7590*, pp. 63–72.

NEA/OECD (1988). *Committee on Radiation Protection and Public Health.* Report of an Expert Group on Gut Transfer Factors. NEA/OECD, Paris.

Rose, E. and Jacobs, H. (1969). Whole-body counter and bioassay results after an acute antimony-126 exposure. *IAEA-SM-119/30*, pp. 269–280.

Tarrant, M. E., Wadley, S. and Woodage, J. J. (1971). The effect of penicillamine on the treatment of experimental schistosomiasis with tartar emetic. *Ann. Trop. Med. Parasitol.* **65**, 233–244.

Thomas, R. G., Felicetti, S. W., Luchino, R. V. and McClellan, R. O. (1973). Retention patterns of antimony in mice following inhalation of particles formed at different temperature. *Proc. Soc. Exp. Biol. Med.* **144**, 544–550.

Van Bruwaene, R., Gerber, G. B., Kirchmann, R. and Colard, K. (1982). Metabolism of antimony-124 in lactating dairy cows. *Health Phys.* **43**, 733–738.

Waitz, J. A., Ober, R. E., Meisenholder, J. E. and Thompson, P. E. (1965). Physiological disposition of antimony after administration of ^{124}Sb-labelled tartar emetic to rat, mice and monkeys, and the effects of tris (*p*-aminophenyl) carbonium pamoate on this distribution. *Bull. W.H.O.* **33**, 537–546.

4. THORIUM

Uptake to blood

(78) *Adults.* The fractional absorption of thorium (Th), administered as $^{234}Th(SO_4)_2$ in the form of mock radium dial paint to six human volunteers (age between 63 and 83 y) gave f_1 values from 10^{-4} to 6×10^{-4}, with a mean value of 2×10^{-4} (Maletskos *et al.*, 1969). Based on these results, an f_1 value of 2×10^{-4} was recommended in *ICRP Publication 30* (ICRP, 1979).

(79) Estimates of the fractional absorption of thorium have also been derived from data on skeleton content, daily dietary intake, inhalation, and urinary excretion. These analyses gave f_1 values from less than 0.001–0.01 (Johnson and Lamothe, 1989; Dang and Sunta, 1990). However, these estimates of f_1 are uncertain because they are based on balance studies from disparate data sources. Unless carefully controlled, such studies do not provide a sound basis for estimating f_1 values.

(80) There have been several experiments to determine the gastrointestinal absorption of inorganic thorium in rats and mice. These experiments gave f_1 values in the range from 5×10^{-5} to 6×10^{-3} for rats (Traikovich, 1970; Pavlovskaya *et al.*, 1971; Sullivan, 1980), about 6×10^{-4} for mice (Sullivan, 1980; Sullivan *et al.*, 1983), and 1×10^{-3} for fasted mice (Larsen *et al.*, 1983).

(81) NEA/OECD (1988) recommended an f_1 value for thorium in food of 0.001 by analogy with plutonium ingested in the tetravalent state, for which an f_1 of 0.001 had been recommended in *ICRP Publication 48* (ICRP, 1986). More recent data on the absorption of neptunium, plutonium, americium, and curium obtained from human volunteer studies (Hunt *et al.*, 1986, 1990; Popplewell *et al.*, 1991, 1994) have provided good support for the use of a lower f_1 value for these elements and a value of 5×10^{-4} has been adopted in *ICRP Publication 67* (ICRP, 1993) for plutonium, americium, and neptunium. Based on the results obtained for thorium by Maletskos *et al.* (1969) and on a comparison with the recent data on the chemically similar element plutonium, an f_1 value of 5×10^{-4} (ICRP, 1993) is adopted here for thorium.

(82) *Children.* With the exception of an experiment of Sullivan *et al.* (1983) with neonatal rats giving an f_1 value of 0.01, there appears to be no age-dependent information available on the gastro-intestinal absorption of thorium. Following the general approach described in Paragraph (8) in the Introduction an f_1 value of 5×10^{-3} for the 3-mo-old infant is adopted here. For children of 1 y and older the f_1 value for adults of 5×10^{-4} is used here.

Distribution and retention

(83) Maletskos *et al.* (1966, 1969) examined the clearance of thorium from blood and its retention and excretion after intravenous injection of ^{234}Th citrate into normal human subjects of age 63–83 y. Detailed measurements were reported for three male and two female subjects (Maletskos *et al.*, 1966). During the first day, thorium cleared from blood with a half-time of a few hours. As an average, about 10% of the injected amount remained in blood after 1 d, 3% after 2 d, 1.5% after 3 d, and 0.3% after 10 d. As indicated by whole-body counting and collection and analysis of excreta, whole-body retention was consistently greater than 90% of the injected amount at 3 weeks after

injection. Cumulative urinary excretion was 4.5–6.1% of the injected amount over the first 5 d after injection, and, as indicated by intermittent measurements, about 2–3% over the next 19 d. The ratio of total urinary excretion to total faecal excretion over the first 5 d was about 12:1 for each of the three males and about 25:1 for both females. External measurements indicated virtually no biological removal from the body during the period from 3–16 weeks after injection. There appeared to be no disproportionate accumulation of thorium in the liver compared with other soft tissues.

(84) Long-term measurements of ^{227}Th or ^{228}Th in the bodies and excreta of occupationally exposed persons suggest a "half-time" for clearance of thorium from the adult human body of at least 10–15 y and probably much more (Rundo, 1964; Newton *et al.*, 1981). Using *in vivo* counting and measurements of the thorium concentration in urine, Dang *et al.* (1992) determined a lower rate of urinary excretion of thorium in occupationally exposed subjects than indicated by the thorium retention model and lung model of *ICRP Publication 30, Part 1* (1979). Sunta *et al.* (1990) estimated that the daily urinary excretion of thorium ranged from 13 to 20% of the serum content in three groups of persons from a thorium processing plant and a group from the general public. In contrast, Hewson and Fardy (1993) estimated a central value of daily urinary excretion of about 2.5% of the serum content in mineral sands workers.

(85) Measurements of thorium isotopes in autopsy samples from non-occupationally exposed subjects (Wrenn *et al.*, 1990; Singh *et al.*, 1983; Ibrahim *et al.*, 1983) indicate that the skeleton probably contains more than three-quarters of the systemic burden during or after chronic exposure to thorium. The contents of the liver and kidneys were variable but typically represented about 2–4% and 0.3–1%, respectively, of the systemic content. These estimates are based on the assumption that muscle, fat, and skin do not accumulate more than 20% of the systemic content, as suggested by data on laboratory animals (Stover *et al.*, 1960; Thomas *et al.*, 1963; Boecker *et al.*, 1963; Traikovich, 1970; Larsen *et al.*, 1984).

(86) There have been a number of studies of the biokinetics of thorium in small mammals, including rats, mice, guinea pigs, and rabbits (Scott *et al.*, 1952; Thomas *et al.*, 1963; Boecker *et al.*, 1963; Traikovich, 1970; Larsen *et al.*, 1984). In many cases, the administration of high concentrations of thorium apparently resulted in colloid formation and high deposition in the reticuloendothelial system or in the tissue into which thorium was introduced (e.g. lung with intratracheal injection, or muscle with intramuscular injection). The results of such studies do not appear to be useful for determining the biokinetics of environmental levels of thorium.

(87) For tracer levels of thorium administered as the citrate to rats, deposition was considerably greater in bone than other systemic tissues (Thomas *et al.*, 1963). Muscle and pelt accounted for about 20% of the systemic activity at 7–54 d post injection.

(88) Boecker *et al.* (1963) found that the level of absorption to blood and the subsequent pattern of distribution and excretion of thorium were not affected by the initial lung content of inhaled thorium in rats. The absorbed activity was deposited mainly in the skeleton. The liver content at 0–40 d was about 15–20% of the skeletal content, and the kidney content during that time was about 3% of the skeletal content. The content of pelt and muscle plus connective tissue was about the same as liver. The urinary to faecal excretion ratio increased gradually to a value of about 0.6–0.7 at 40–50 d post inhalation.

(89) Larsen *et al.* (1984) observed a similar distribution of thorium in mice after gastrointestinal absorption and intravenous injection. At 3 d after injection, about 90%

of the systemic content was found in the skeleton, 6% in liver, 4% in kidneys, and 0.1% in reproductive organs. A urinary to faecal excretion ratio of 16:1 was observed.

(90) Stover *et al.* (1960) studied the biokinetics of ^{228}Th in adult beagle dogs over a 1300 d period following intravenous injection as the citrate complex. About 88% of the injected amount was retained at 3 weeks, 80% at 1 y, and 70% at 3.5 y. The urinary excretion rate was about 4 times the faecal excretion rate in the first few weeks, but the urinary-to-faecal excretion ratio gradually decreased and was close to 1 at 2.5 y after injection. About 70% of injected thorium deposited in the skeleton, about 5% deposited in the liver, and about 3% deposited in the kidneys. At times remote from injection ($\geq$ 100 d), about 80% of retained thorium was in the skeleton and about 20% was widely distributed in soft tissues, with significant amounts in the liver and kidneys. There was little if any decline in the thorium content of compact bone over 1300 d or in trabecular bone over 0–800 d, but there was a noticeable decline in activity in trabecular bone over 800–1300 d. The thorium content of the liver, kidneys, and spleen declined considerably in the first several months after injection but showed little or no decrease thereafter. Retention of thorium in the kidneys and/or its rate of urinary excretion at times remote from injection may have been affected by radiation damage at high dosage levels (Stover *et al.*, 1960). Thorium is distributed throughout the kidney tissue but is highly concentrated in the arterioles and glomeruli (Jee *et al.*, 1962).

(91) Comparison of the organ distributions of thorium isotopes in humans and beagles exposed only to environmental levels indicate that the beagle may be a reasonable experimental model for man with regard to the long-term distribution of thorium in the body (Singh *et al.*, 1988). There are also broad similarities in the patterns of distribution and excretion of injected thorium at early times in human subjects (Maletskos *et al.*, 1966, 1969) and beagles (Stover *et al.*, 1960).

(92) The metabolic behaviour of thorium in the body appears to be similar in many respects to that of the more frequently studied element, plutonium (Hamilton, 1947; Stover *et al.*, 1960; Jee *et al.*, 1962; Singh *et al.*, 1983). For example, the pattern of distribution of thorium in bone is similar to that of plutonium (Jee *et al.*, 1962; Jee, 1964). Both elements deposit on bone surfaces, where they are tenaciously retained until buried in bone volume or removed to blood and/or bone marrow by bone restructuring processes (Jee *et al.*, 1962; Jee, 1964). Also, total-body retention in beagles is roughly the same for thorium and plutonium over a period of years (Stover *et al.*, 1960). In humans chronically exposed to environmental levels of thorium and plutonium, the content of the gonads as a percentage of the systemic content appears to be roughly the same for the two elements (Singh *et al.*, 1983).

(93) There are, however, important differences in the biokinetics of these two elements. For example, it appears that thorium accumulates in the liver to a much smaller extent than plutonium (at least, in dogs and humans), has higher urinary clearance and perhaps lower faecal clearance from blood than plutonium, and accumulates to a greater extent than plutonium in the kidneys and blood vessels of dogs (Stover *et al.*, 1960; Jee *et al.*, 1962; Singh *et al.*, 1983).

The structure of the biokinetic model for thorium

(94) The retention model for thorium given in *ICRP Publication 30* (ICRP, 1979), was based directly on data for beagles intravenously injected with ^{228}Th (Stover *et al.*, 1960). According to that model, thorium leaves the transfer compartment (circulation) with a half-time of 0.5 d. Of the activity leaving this compartment, 70% is assigned to

bone surfaces, from which it is removed with a biological half-time of 8000 d; 4% is assigned to the liver, from which it is removed with a half-time of 700 d; 16% is assumed to be uniformly distributed in the rest of the body and removed with a half-time of 700 d; and the remaining 10% is assumed to be removed directly by excretion.

(95) The representation of the skeleton as a well-mixed pool is not compatible with what is now known about the biological behaviour of actinides. Moreover, it appears that the simple structure of the model for thorium given in *ICRP Publication 30* cannot be used to reproduce the long-term patterns of retention in major depots. For example, in contrast to the structure of the model of *ICRP Publication 30*, the data for beagles suggest potential differences in the retention of thorium in compact and trabecular bone after a few years (as discussed above) and indicate that the contents of soft tissues do not decline exponentially over an extended period.

(96) Based on these considerations and the apparent similarities in the behaviour of thorium and plutonium with regard to the skeleton, the more detailed model structure developed for plutonium and other "bone-surface-seeking" elements (Fig. 4.1, ICRP, 1993) is applied here to thorium. Because some transfer rates in the biokinetic model for thorium are equated with bone formation rates which are expected to remain elevated during the early part of the third decade of life, the dose coefficients for the adult are based on equivalent dose rates received over a 50 y period following acute intake at age 25 y.

(97) In this model, blood is treated as a uniformly mixed pool. Compartment ST0 is a soft tissue pool that includes the extracellular fluids and exchanges material with blood over a period of hours or days. Compartment ST0 is used in the model for two purposes: to depict an early build-up and decline of material in soft tissues, and to account for early feedback of material to blood. Thus, compartment ST0 represents an integral part of the early circulation of material. In the summary of parameter values below, deposition fractions for compartments other than ST0 are given in terms of "activity leaving the circulation" and refer to the division of material among compartments other than ST0.

(98) The skeleton is divided into cortical and trabecular fractions, and each of these is subdivided into bone surfaces, bone volume, and bone marrow. Activity entering the skeleton is assigned initially to bone surfaces and is subsequently transferred to bone marrow by bone resorption or to bone volume by bone formation. Activity in bone volume is transferred to bone marrow by bone resorption. Activity moves from bone marrow to blood over a period of months and is subsequently redistributed in the same pattern as the original input to blood.

(99) The liver is viewed as consisting of two compartments, Liver 1 and Liver 2. Liver 2 represents relatively tenacious retention ($T_{1/2} > 1$ y) in the liver; it is defined on a kinetic rather than a biological basis but may be associated largely with the reticuloendothelial system. Thorium entering the liver is assumed to deposit in Liver 1. Fractions of the activity leaving Liver 1 are assigned to blood, to Liver 2, and to the GI contents (via biliary secretion). Thorium leaving Liver 2 is assumed to return to blood.

(100) The kidneys are assumed to consist of two compartments, one that loses thorium to urine and another that returns activity to blood. The "urinary bladder contents" is considered as a separate pool that receives all material destined for urinary excretion.

(101) Soft tissue compartments ST1 and ST2 are used to represent intermediate-term retention (up to a few years) and tenacious retention (many years), respectively, in the other soft tissues.

(102) Movement of material in the body is depicted as a system of first-order processes. Parameter values are expressed as transfer rates between compartments, but most of the derived transfer rates are secondary values calculated from selected "deposition fractions" and "removal half-times". The concepts of "deposition fraction" and "removal half-time" are commonly used in connection with non-recycling biokinetic models but require some explanation with regard to the recycling models used in this report.

(103) The "removal half-time" from a compartment refers to the biological half-time that would be observed if there were no recycling to that compartment. This will generally differ from the apparent (or net, or externally viewed) half-time that may be estimated at any given time in the presence of recycling. For example, the removal half-time of thorium from compartment ST0 to blood is assumed to be 1.5 d (that is, the transfer rate from ST0 to blood is $\ln(2)/1.5 = 0.462\ d^{-1}$), but it would require more than 1.5 d for the contents of ST0 to be reduced by 50% because of a continuous feed from blood. Similarly, the removal half-time of thorium from blood (by all pathways combined) is assumed to be 0.25 d, but the disappearance curve from blood at 2–4 d suggests a much longer half-time because flow to blood from ST0 and other compartments would largely offset outflow from blood during that period.

(104) Transfer rates from blood to various compartments are based on rounded "deposition fractions", which provide a convenient way to describe the initial distribution of activity leaving the circulation. For example, the transfer rate of thorium from blood to Liver 1 in the adult is based on the estimates that 30% of activity leaving blood is deposited in the rapid-turnover soft tissue compartment ST0, and 5% of thorium leaving the circulation (i.e. leaving blood but not depositing in ST0) deposits in Liver 1. Since the total removal rate from blood is $\ln(2)/0.25 = 2.77\ d^{-1}$ and 30% of this goes to ST0, the transfer rate to all other tissues and excreta is $0.7 \times 2.77 = 1.94\ d^{-1}$. Therefore, the transfer rate of thorium from blood to Liver 1 is $0.05 \times 1.94 = 0.097\ d^{-1}$.

Parameter values for adults

(105) The studies described above demonstrate that the behaviour of thorium in the body can be influenced by the mass concentration. In choosing model parameters it is important to select values that are appropriate for environmental concentrations of thorium. For workers exposed occupationally to high concentrations of thorium alternative parameter values may be needed.

(106) The model of *ICRP Publication 56* (ICRP, 1993) for bone-surface-seeking radionuclides is generic in the sense that a common model structure is applied to all bone-surface-seeking elements, and the model involves some generic transfer rates associated with bone restructuring. Many of the transfer rates, however, must be developed on an element-by-element basis. As far as is practical, these transfer rates were derived from thorium-specific data described earlier. Where data on thorium were not available, transfer rates developed for plutonium (ICRP, 1993) were applied. Except where indicated to the contrary, parameter values are independent of age.

(107) *Clearance from the systemic circulation:* Data on clearance of thorium from blood in human subjects over the first 10 d after injection (Maletskos *et al.*, 1966) can be reproduced reasonably well by treating blood as a uniformly mixed pool that loses activity with a half-time of 0.25 d and postulating a soft tissue compartment (ST0) that receives 30% of activity leaving blood and returns activity to blood with a half-time of

1.5 d. The assumption that ST0 receives 30% of the instantaneous outflow from blood is a default assumption applied in this modelling scheme when there is little or no direct information on the early build-up and decline of material in soft tissues. Removal half-times of 0.25 d from blood and 1.5 d from compartment ST0 were chosen to yield a reasonable approximation of the blood clearance curve for injected thorium in human subjects.

(108) *Skeleton:* In view of thorium injection data for beagles and autopsy data on humans chronically exposed to this element, the skeleton must be assigned a large portion of thorium leaving the circulation. In this model it is assumed that 70% of thorium leaving the circulation deposits on bone surfaces, with half depositing on cortical surfaces and half on trabecular surfaces. The model for translocation of skeletal deposits of thorium is the same as that for plutonium, americium, and neptunium given in *ICRP Publications 56* and *67*. According to that model, the rate of translocation of skeletal deposits is controlled by bone restructuring processes. As described in *ICRP Publication 67*, the transfer rates for the adult apply to ages ≥25 y because bone formation rates are expected to remain elevated during the early part of the third decade of life. The transfer rate from compact or trabecular bone surfaces to the corresponding bone marrow compartment is assumed to equal the rate at which that type of bone surface is resorbed. The transfer rate from bone surfaces to bone volume is set equal to one-half the surface formation rate. The transfer rate from compact or trabecular bone volume to the corresponding bone marrow compartment is set equal to the rate at which that type of bone volume is resorbed. The bone formation rates are the same as those used in *ICRP Publications 56* and *67* and, as in those documents, a common value is assumed for bone formation and resorption and is applied both to surface and volume remodelling. The removal half-time from bone marrow to blood is assumed to be 0.25 y, the same value used for plutonium, americium, and neptunium in *ICRP Publication 67*. This value was originally derived for plutonium and was based on data for beagles, but model predictions based on this removal rate are generally consistent with recent measurements of plutonium and americium in bone and bone marrow of occupationally exposed persons (McInroy and Kathren, 1990). All material leaving bone marrow is assumed to enter blood.

(109) *Liver:* The model for the liver is based on the following sources of information: the early pattern of retention of thorium in the beagle liver; the pattern of intermediate-term decline in the thorium content of the beagle liver; the decreasing urinary-to-faecal excretion ratio, which is assumed to be due to continuing biliary secretion; and data on the relative liver content, as a portion of the systemic content in human subjects chronically exposed to thorium. It is assumed that 5% of activity leaving the circulation deposits in Liver 1. The removal half-time from Liver 1 is 1 y, which is a default value used in the modelling scheme for transuranium elements (ICRP, 1993). Of activity leaving Liver 1, 50% is assigned to Liver 2, 25% to blood, and 25% to the GI tract contents. The long-term compartment, Liver 2, is assumed to lose activity to blood with a biological half-time of 9 y, a value derived for plutonium (ICRP, 1993). The model predicts that during or after chronic exposure to thorium, the liver (primarily Liver 2) contains 3–4% of the total systemic content, in reasonable agreement with data on chronically exposed humans. The model predicts higher retention in the human liver than was observed in beagles during the period 200–1000 d after injection, but limited data suggest that the human liver may accumulate a higher portion of systemic thorium than does the beagle liver (Singh *et al.*, 1988).

(110) *Soft tissue compartments ST1 and ST2:* The model for intermediate and long-term retention in other soft tissues is based on data and estimates derived for laboratory animals (Stover *et al.*, 1960; Thomas *et al.*, 1963; Boecker *et al.*, 1963; Traikovich, 1970; Larsen *et al.*, 1984) and analogy with plutonium and americium (ICRP, 1993). It is assumed that 2% of activity leaving the circulation deposits in a soft tissue compartment, designated as ST2, with nearly permanent retention. The percentage left over after all other deposition fractions in the model have been chosen, amounting to 12.5% of thorium leaving the circulation, is assigned to the intermediate-turnover soft tissue compartment ST1. By analogy with plutonium, the removal half-times from ST1 and ST2 to blood are assumed to be 2 y and 100 y, respectively. The model predicts that the thorium content of other soft tissues (including ST0) peaks at about 30% of the injected amount at 1 d, declines to about 20% at 5 d and 15% at 10 d, and then declines slowly to about 10% at 1000 d and 6% at 10,000 d. These estimates are broadly consistent with the limited data on laboratory animals, although these data extend at most to a few years after exposure (Stover *et al.*, 1960; Thomas *et al.*, 1963; Boecker *et al.*, 1963; Traikovich, 1970; Larsen *et al.*, 1984). The model predicts that the other soft tissues contain about 8% of the systemic content during or after long-term chronic exposure, in reasonable agreement with human data on the more frequently studied actinides, plutonium and americium (McInroy *et al.*, 1989).

(111) *Gonads:* The model for uptake and retention of thorium in gonads is the same as that applied to plutonium, americium, and neptunium in *ICRP Publication 67.* This model is based on recommendations given in *ICRP Publication 48* (ICRP, 1986), except that a finite retention time in the gonads was assigned. It is assumed that deposition in the gonads, expressed as a percentage of thorium leaving the circulation, is 0.001% g^{-1} gonadal tissue. This yields a deposition of 0.035% of thorium leaving the circulation in the 35 g testes of the reference adult male and 0.011% in the 11 g ovaries of the reference adult female (ICRP, 1975). The removal half-time from gonads to blood is assumed to be 10 y; this value is reasonably consistent with the assumption in *ICRP Publication 48* of permanent retention in gonadal tissue due to consideration in the present model of continuous gonadal uptake of activity lost from other tissues. The model predicts that the testes contain about 0.05% of the systemic content of thorium during long-term chronic exposure, compared with values of 0.01–0.09% determined in chronically exposed human subjects (Singh *et al.*, 1983).

(112) *Excretion pathways:* Parameter values for urinary excretion and renal retention of thorium were chosen for consistency with urinary excretion data on human subjects injected with thorium for experimental purposes, data on the relative organ contents of human subjects chronically exposed to thorium, and kidney retention data from beagles injected with thorium. It is assumed that 5.5% of activity leaving the circulation goes directly to the urinary bladder and that another 3.5% deposits in the renal tubules and is subsequently released to the urinary bladder contents with a half-time of 15 d. These parameter values yield good agreement between model predictions and early urinary excretion data for humans injected with thorium. For a scenario of chronic exposure (more specifically, constant absorption to blood from external sources), the model predicts that the daily urinary excretion of thorium is about 17% of the blood content. Recently published data give rather disparate results for the clearance of thorium to the urine. Sunta *et al.* (1990) found that the daily urinary excretion of thorium as a percentage of the serum pool was in the range 13–20% in human subjects. In contrast, Hewson and Fardy (1993) determined a geometric mean value of 2.5% for mineral sands workers.

(113) To account for the relatively high concentration of thorium in the kidneys of laboratory animals at times remote from intake and in the kidneys of chronically exposed humans, it is assumed that another 1% of thorium leaving the circulation deposits in other kidney tissues, from which it is returned to blood with a half-time of 5 y. With these parameter values, the model predicts that the kidneys contain about 0.7% of the systemic content of thorium under conditions of chronic intake, which is consistent with autopsy data on human subjects. The predicted pattern of decline of the kidney content as a function of time after injection is consistent with thorium-injection data on beagles.

(114) Data on humans and beagles (Stover *et al.*, 1960; Maletskos *et al.*, 1966, 1969) indicate fairly rapid movement of a small amount of thorium to faeces, and the longer-term study on beagles also indicates a gradual decrease with time in the urinary-to-faecal excretion ratio. In this model, early faecal excretion is accounted for primarily by secretion directly from blood into the GI contents, and longer-term faecal excretion primarily by biliary secretion from the liver (Liver 1 in this case) into the GI contents. Specifically, it is assumed that 0.5% of thorium leaving the circulation is secreted into the upper large intestine and 25% of that leaving Liver 1 transfers via biliary secretion into the small intestine. Thorium entering the small intestine is assumed to be available for reabsorption to blood, with fractional absorption being the same as for directly ingested thorium. The model predicts that the ratio of cumulative urinary excretion to cumulative faecal excretion is about 15 after 5 d, which is consistent with the ratios of about 12 and 25 determined for male and female subjects, respectively. The daily urinary-to-faecal excretion ratio is predicted to decline to about 1.3 at 0.5 y after injection and then to increase gradually to about 5 over a period of years as Liver 1 becomes depleted of its initial deposit.

(115) *Children.* There is a paucity of age-specific information on the biokinetics of thorium. Since immature animals (including humans) generally have a higher skeletal deposition of bone-seeking radioelements than mature adults, it is arbitrarily assumed that 80% of injected thorium deposits on bone surfaces (40% on trabecular surfaces and 40% on cortical surfaces) in persons of age ≤ 15 y, compared with 70% in the adult.

(116) No age-specific data were found for the retention of thorium in the liver, kidneys or in the other soft tissues. The transfer rates from these compartments are assumed to be independent of age.

(117) The age-specific transfer rate from bone surfaces to bone volume is equal to the surface formation rate in persons of age 0–15 y. The age-specific transfer rate from compact or trabecular bone volume to the corresponding bone marrow compartment is set equal to the rate at which that type of bone volume is resorbed. The age-specific bone formation rates are the same as those assumed in *ICRP Publications 56* and *67* and, as in those documents, a common value is used for bone formation and resorption and is applied both to surface and volume remodelling.

(118) For children, deposition in gonadal tissue is assumed to give twice the concentration in adults, or 0.002% g^{-1} gonadal tissue; this is intended as a cautious assumption with regard to estimated dose equivalent to gonadal tissues, as judged from comparative gonadal uptake of plutonium in immature beagles with that in mature beagles (Miller *et al.*, 1989). Based on data given in the ICRP Reference Man document (ICRP, 1975), the weight of the testes is assumed to be 1 g in infants, 1.5 g at age 1 y, 1.7 g at age 5 y, 2 g at age 10 y, and 16 g at age 15 y; the weight of the ovaries is assumed to be 0.6 g in infants, 0.8 g at age 1 y, 2.0 g at age 5 y, 3.5 g at age 10 y, and 6 g at age 15 y. The removal half-time from gonads to blood is assumed to be 10 y for all age groups.

(119) Deposition fractions in the soft tissues and excretion pathways assigned to adults are reduced by one-third for children to accommodate higher skeletal deposition.

Behaviour of decay products

(120) The treatment of decay products is similar to that described in Chapter 13 of *ICRP Publication 67* for radium, but the updated model for thorium described in the present document is applied to isotopes of thorium, actinium, and protoactinium produced *in vivo*. For treatment of radium and lead isotopes produced *in vivo*, the bone volume compartments shown in Fig. 4.1 must be divided into exchangeable and non-exchangeable bone volume and blood must be divided into plasma and red blood cells (see Fig. 5.1, which shows the generic model structure for calcium-like elements). For treatment of thorium, actinium, or protoactinium, the blood compartment shown in Fig. 4.1 is identified with blood plasma and the bone volume compartments shown in that figure are identified with the corresponding non-exchangeable bone volume compartments in Fig. 5.1. Thorium, actinium, or protoactinium produced in exchangeable bone volume is assumed to move immediately (implemented using a transfer rate of 1000 d^{-1}) to the corresponding non-exchangeable bone volume compartment. For treatment of radium isotopes produced in compartments included in the present thorium model but not addressed in the radium model in *ICRP Publication 67* (Liver 2, testes, ovaries, bone marrow compartments, and kidney compartments), the assumption is made that radium follows the direction of movement of thorium but

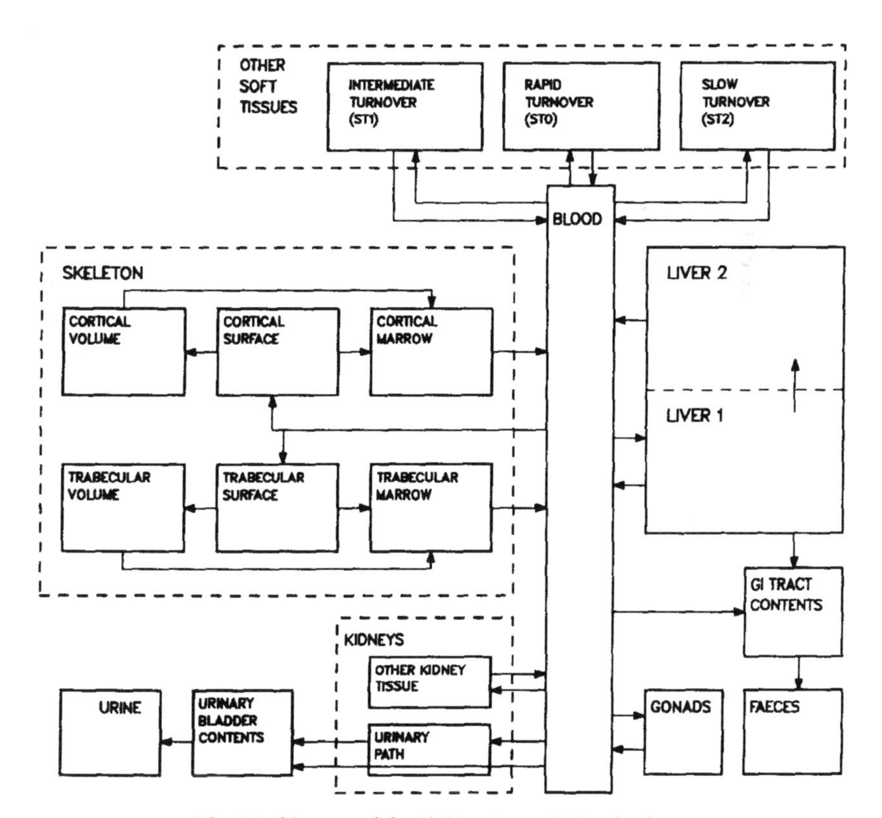

Fig. 4.1. Diagram of the biokinetic model for thorium.

leaves Liver 2 with a removal half-time of 50 d (the half-time assigned to Liver 1) and leaves gonads, bone marrow compartments, and kidney compartments with a removal half-time of 1 d (the half-time assigned to the intermediate turnover in other soft tissues, ST1). Lead isotopes produced in compartments not addressed in the lead model of *ICRP Publication 67* (gonads and bone marrow) are also assumed to be removed to plasma with a half-time of 1 d. The models for polonium and bismuth produced *in vivo* are given in the section on lead in *ICRP Publication 67* (ICRP, 1993) and the model for radon is given in the section on radium. Thallium-208 ($T_{1/2} = 3.08$ min) is assumed to decay at its point of origin.

(121) Figures 4.2 and 4.3 give estimated thorium contents of bone and liver in infants, 10-y-old children, and adults as a function of time after injection, based on the parameter values given in Table 4.-1.

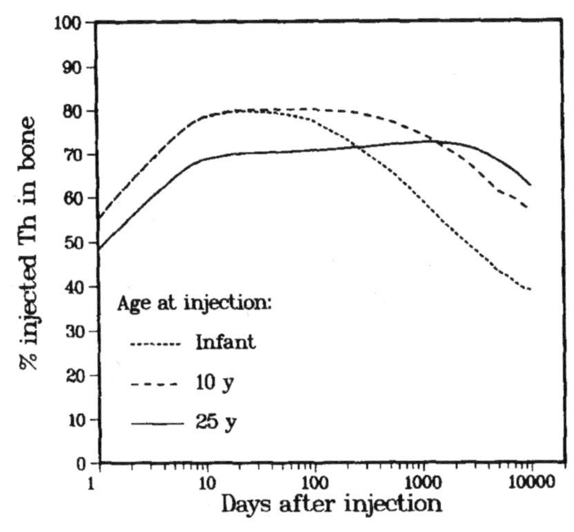

Fig. 4.2. Model predictions of the thorium content of bone as a function of time after injection into blood for different ages.

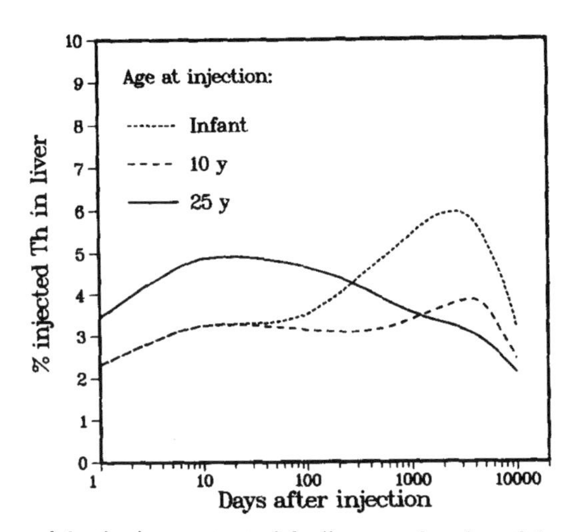

Fig. 4.3. Model predictions of the thorium content of the liver as a function of time after injection into blood for different ages.

Dose coefficients

(122) Dose coefficients derived from the biokinetic data summarised in Table 4.-1 are given in Tables 4.-2. to 4.-5.

Table 4.-1. Age-specific transfer rates (d^{-1}) for thorium model

	Age					
	3 mo	1 y	5 y	10 y	15 y	Adult
Blood to Liver 1	6.470E-02	6.470E-02	6.470E-02	6.470E-02	6.470E-02	9.700E-02
Blood to Cortical surface	7.763E-01	7.763E-01	7.763E-01	7.763E-01	7.763E-01	6.793E-01
Blood to Trabecular surface	7.763E-01	7.763E-01	7.763E-01	7.763E-01	7.763E-01	6.793E-01
Blood to Urinary bladder	7.110E-02	7.110E-02	7.110E-02	7.110E-02	7.110E-02	1.067E-01
Blood to Kidneys (Urinary path)	4.530E-02	4.530E-02	4.530E-02	4.530E-02	4.530E-02	6.790E-02
Blood to Other kidney tissue	1.290E-02	1.290E-02	1.290E-02	1.290E-02	1.290E-02	1.940E-02
Blood to Large intestine	6.470E-03	6.470E-03	6.470E-03	6.470E-03	6.470E-03	9.700E-03
Blood to Testes	3.900E-05	5.800E-05	6.600E-05	7.700E-05	6.200E-04	6.800E-04
Blood to Ovaries	2.300E-05	3.000E-05	7.600E-05	1.300E-04	2.300E-04	2.100E-04
Blood to ST0	8.320E-01	8.320E-01	8.320E-01	8.320E-01	8.320E-01	8.320E-01
Blood to ST1	1.620E-01	1.620E-01	1.620E-01	1.620E-01	1.620E-01	2.430E-01
Blood to ST2	2.590E-02	2.590E-02	2.590E-02	2.590E-02	2.590E-02	3.880E-02
ST0 to Blood	4.620E-01	4.620E-01	4.620E-01	4.620E-01	4.620E-01	4.620E-01
Kidneys (Urinary path) to Bladder	4.620E-02	4.620E-02	4.620E-02	4.620E-02	4.620E-02	4.620E-02
Other kidney tissue to Blood	3.800E-04	3.800E-04	3.800E-04	3.800E-04	3.800E-04	3.800E-04
ST1 to Blood	9.500E-04	9.500E-04	9.500E-04	9.500E-04	9.500E-04	9.500E-04
ST2 to Blood	1.900E-05	1.900E-05	1.900E-05	1.900E-05	1.900E-05	1.900E-05
Trabecular surface to Volume	8.220E-03	2.880E-03	1.810E-03	1.320E-03	9.590E-04	2.470E-04
Trabecular surface to Marrow	8.220E-03	2.880E-03	1.810E-03	1.320E-03	9.590E-04	4.930E-04
Cortical surface to Volume	8.220E-03	2.880E-03	1.530E-03	9.040E-04	5.210E-04	4.110E-05
Cortical surface to Marrow	8.220E-03	2.880E-03	1.530E-03	9.040E-04	5.210E-04	8.210E-05
Trabecular volume to Marrow	8.220E-03	2.880E-03	1.810E-03	1.320E-03	9.590E-04	4.930E-04
Cortical volume to Marrow	8.220E-03	2.880E-03	1.530E-03	9.040E-04	5.210E-04	8.210E-05
Cort/trab bone marrow to Blood	7.600E-03	7.600E-03	7.600E-03	7.600E-03	7.600E-03	7.600E-03
Liver 1 to Liver 2	9.500E-04	9.500E-04	9.500E-04	9.500E-04	9.500E-04	9.500E-04
Liver 1 to Small intestine	4.750E-04	4.750E-04	4.750E-04	4.750E-04	4.750E-04	4.750E-04
Liver 1 to Blood	4.750E-04	4.750E-04	4.750E-04	4.750E-04	4.750E-04	4.750E-04
Liver 2 to Blood	2.110E-04	2.110E-04	2.110E-04	2.110E-04	2.110E-04	2.110E-04
Gonads to Blood	1.900E-04	1.900E-04	1.900E-04	1.900E-04	1.900E-04	1.900E-04
f_1	5.000E-03	5.000E-04	5.000E-04	5.000E-04	5.000E-04	5.000E-04

Parameters are given to sufficient precision for calculational purposes. This may be more precise than the biological data would support.

Table 4.-2.

Ingestion Dose Coefficients: Committed Equivalent and Effective Doses per Unit Intake (Sv/Bq) for Th-228 (T1/2 = 1.9131 y)*

Age at intake	3 Months	1 Year	5 Years	10 Years	15 Years	Adult
Adrenals	4.0E-07	3.5E-08	1.9E-08	1.0E-08	6.7E-09	7.2E-09
Bladder Wall	4.0E-07	3.5E-08	1.9E-08	1.1E-08	6.8E-09	7.3E-09
Bone Surfaces	8.8E-05	8.4E-06	6.3E-06	4.3E-06	3.3E-06	2.5E-06
Brain	4.0E-07	3.5E-08	1.9E-08	1.0E-08	6.7E-09	7.2E-09
Breast	4.0E-07	3.4E-08	1.9E-08	1.0E-08	6.6E-09	7.2E-09
GI-Tract						
St Wall	4.1E-07	4.4E-08	2.3E-08	1.3E-08	8.4E-09	8.5E-09
SI Wall	4.4E-07	6.2E-08	3.2E-08	1.9E-08	1.1E-08	1.1E-08
ULI Wall	8.3E-07	2.7E-07	1.4E-07	8.0E-08	4.5E-08	4.0E-08
LLI Wall	2.1E-06	1.0E-06	5.1E-07	3.0E-07	1.7E-07	1.5E-07
Kidneys	3.1E-06	2.5E-07	1.5E-07	9.4E-08	6.7E-08	6.4E-08
Liver	5.6E-06	4.5E-07	2.5E-07	1.5E-07	9.4E-08	1.0E-07
Lungs	4.0E-07	3.5E-08	1.9E-08	1.0E-08	6.7E-09	7.2E-09
Muscle	4.0E-07	3.5E-08	1.9E-08	1.0E-08	6.7E-09	7.2E-09
Ovaries	6.7E-07	6.4E-08	5.3E-08	3.6E-08	2.4E-08	2.0E-08
Pancreas	4.0E-07	3.5E-08	1.9E-08	1.0E-08	6.7E-09	7.2E-09
Red Marrow	1.7E-05	1.3E-06	7.1E-07	4.2E-07	2.8E-07	1.9E-07
Skin	4.0E-07	3.5E-08	1.9E-08	1.0E-08	6.6E-09	7.2E-09
Spleen	4.0E-07	3.5E-08	1.9E-08	1.0E-08	6.7E-09	7.2E-09
Testes	7.8E-07	7.6E-08	5.5E-08	4.3E-08	3.1E-08	2.0E-08
Thymus	4.0E-07	3.4E-08	1.9E-08	1.0E-08	6.6E-09	7.2E-09
Thyroid	4.0E-07	3.4E-08	1.9E-08	1.0E-08	6.6E-09	7.2E-09
Uterus	4.0E-07	3.5E-08	1.9E-08	1.1E-08	6.8E-09	7.3E-09
Remainder	4.5E-07	3.9E-08	2.1E-08	1.2E-08	7.4E-09	7.8E-09
Effective Dose	3.7E-06	3.7E-07	2.2E-07	1.4E-07	9.3E-08	7.2E-08

GI-Tract Gastrointestinal Tract
St Stomach
SI Small Intestine
ULI Upper Large Intestine
LLI Lower Large Intestine

* In the biokinetic model for Th parameter values for the adult apply to ages > 25 y. For radioisotopes of this element the dose coefficients for the adult are based on the 50-y integrated doses following an acute intake at age 25 y.

Table 4.-3.

Ingestion Dose Coefficients: Committed Equivalent and Effective Doses per Unit Intake (Sv/Bq) for Th-230 (T1/2 = 7.7E4 y)*

Age at intake	3 Months	1 Year	5 Years	10 Years	15 Years	Adult
Adrenals	4.6E-07	4.1E-08	2.8E-08	2.0E-08	1.5E-08	1.4E-08
Bladder Wall	4.6E-07	4.1E-08	2.8E-08	2.0E-08	1.5E-08	1.4E-08
Bone Surfaces	1.2E-04	1.3E-05	1.2E-05	1.1E-05	1.1E-05	1.2E-05
Brain	4.6E-07	4.1E-08	2.8E-08	2.0E-08	1.5E-08	1.4E-08
Breast	4.6E-07	4.1E-08	2.8E-08	2.0E-08	1.5E-08	1.4E-08
GI-Tract						
St Wall	4.7E-07	4.8E-08	3.1E-08	2.2E-08	1.7E-08	1.5E-08
SI Wall	4.9E-07	6.1E-08	3.8E-08	2.6E-08	1.9E-08	1.6E-08
ULI Wall	6.6E-07	1.7E-07	9.1E-08	5.7E-08	3.6E-08	3.0E-08
LLI Wall	1.0E-06	4.0E-07	2.1E-07	1.3E-07	7.6E-08	6.2E-08
Kidneys	5.4E-06	4.8E-07	3.4E-07	2.5E-07	2.0E-07	1.9E-07
Liver	4.8E-06	4.2E-07	3.0E-07	2.2E-07	1.7E-07	1.6E-07
Lungs	4.6E-07	4.1E-08	2.8E-08	2.0E-08	1.5E-08	1.4E-08
Muscle	4.6E-07	4.1E-08	2.8E-08	2.0E-08	1.5E-08	1.4E-08
Ovaries	2.0E-06	2.0E-07	1.9E-07	1.5E-07	1.3E-07	9.9E-08
Pancreas	4.6E-07	4.1E-08	2.8E-08	2.0E-08	1.5E-08	1.4E-08
Red Marrow	1.5E-05	1.3E-06	8.7E-07	5.9E-07	4.7E-07	4.2E-07
Skin	4.6E-07	4.1E-08	2.8E-08	2.0E-08	1.5E-08	1.4E-08
Spleen	4.6E-07	4.1E-08	2.8E-08	2.0E-08	1.5E-08	1.4E-08
Testes	2.2E-06	2.2E-07	1.7E-07	1.4E-07	1.3E-07	1.0E-07
Thymus	4.6E-07	4.1E-08	2.8E-08	2.0E-08	1.5E-08	1.4E-08
Thyroid	4.6E-07	4.1E-08	2.8E-08	2.0E-08	1.5E-08	1.4E-08
Uterus	4.6E-07	4.1E-08	2.8E-08	2.0E-08	1.5E-08	1.4E-08
Remainder	5.3E-07	4.7E-08	3.2E-08	2.2E-08	1.7E-08	1.5E-08
Effective Dose	4.1E-06	4.1E-07	3.1E-07	2.5E-07	2.2E-07	2.1E-07

GI-Tract Gastrointestinal Tract
St Stomach
SI Small Intestine
ULI Upper Large Intestine
LLI Lower Large Intestine

* In the biokinetic model for Th parameter values for the adult apply to ages > 25 y. For radioisotopes of this element the dose coefficients for the adult are based on the 50-y integrated doses following an acute intake at age 25 y.

Table 4.-4.

Ingestion Dose Coefficients: Committed Equivalent and Effective Doses per Unit Intake (Sv/Bq) for Th-232 (T1/2 = 1.405E10 y)*

Age at intake	3 Months	1 Year	5 Years	10 Years	15 Years	Adult
Adrenals	1.0E-06	9.4E-08	7.0E-08	5.3E-08	4.1E-08	3.6E-08
Bladder Wall	1.0E-06	9.4E-08	7.0E-08	5.3E-08	4.1E-08	3.6E-08
Bone Surfaces	1.3E-04	1.3E-05	1.3E-05	1.2E-05	1.2E-05	1.2E-05
Brain	1.0E-06	9.4E-08	7.0E-08	5.3E-08	4.1E-08	3.6E-08
Breast	1.0E-06	9.4E-08	7.0E-08	5.2E-08	4.1E-08	3.6E-08
GI-Tract						
St Wall	1.0E-06	1.0E-07	7.3E-08	5.4E-08	4.3E-08	3.7E-08
SI Wall	1.1E-06	1.1E-07	7.9E-08	5.8E-08	4.4E-08	3.8E-08
ULI Wall	1.2E-06	2.0E-07	1.3E-07	8.6E-08	6.0E-08	5.1E-08
LLI Wall	1.5E-06	4.1E-07	2.3E-07	1.5E-07	9.6E-08	8.0E-08
Kidneys	5.1E-06	4.6E-07	3.3E-07	2.5E-07	2.0E-07	1.8E-07
Liver	5.1E-06	4.6E-07	3.4E-07	2.5E-07	2.1E-07	1.8E-07
Lungs	1.0E-06	9.4E-08	7.0E-08	5.3E-08	4.1E-08	3.6E-08
Muscle	1.0E-06	9.4E-08	7.0E-08	5.3E-08	4.1E-08	3.6E-08
Ovaries	2.2E-06	2.2E-07	2.0E-07	1.6E-07	1.4E-07	1.0E-07
Pancreas	1.0E-06	9.4E-08	7.0E-08	5.3E-08	4.1E-08	3.6E-08
Red Marrow	1.5E-05	1.4E-06	9.4E-07	6.8E-07	5.4E-07	4.6E-07
Skin	1.0E-06	9.4E-08	7.0E-08	5.2E-08	4.1E-08	3.6E-08
Spleen	1.0E-06	9.4E-08	7.1E-08	5.3E-08	4.2E-08	3.6E-08
Testes	2.4E-06	2.4E-07	1.9E-07	1.5E-07	1.4E-07	1.0E-07
Thymus	1.0E-06	9.4E-08	7.0E-08	5.2E-08	4.1E-08	3.6E-08
Thyroid	1.0E-06	9.4E-08	7.0E-08	5.3E-08	4.1E-08	3.6E-08
Uterus	1.0E-06	9.4E-08	7.0E-08	5.2E-08	4.1E-08	3.6E-08
Remainder	1.1E-06	9.9E-08	7.3E-08	5.5E-08	4.3E-08	3.7E-08
Effective Dose	4.6E-06	4.5E-07	3.5E-07	2.9E-07	2.5E-07	2.3E-07

GI-Tract Gastrointestinal Tract
St Stomach
SI Small Intestine
ULI Upper Large Intestine
LLI Lower Large Intestine

* In the biokinetic model for Th parameter values for the adult apply to ages > 25 y. For radioisotopes of this element the dose coefficients for the adult are based on the 50-y integrated doses following an acute intake at age 25 y.

Table 4.-5.

Ingestion Dose Coefficients: Committed Equivalent and Effective Doses per Unit Intake (Sv/Bq) for Th-234 (T1/2 = 24.10 d)*

Age at intake	3 Months	1 Year	5 Years	10 Years	15 Years	Adult
Adrenals	6.6E-11	1.4E-11	7.5E-12	4.3E-12	2.6E-12	2.1E-12
Bladder Wall	1.2E-10	5.3E-11	2.9E-11	2.0E-11	1.2E-11	9.9E-12
Bone Surfaces	3.6E-09	2.6E-10	1.7E-10	9.4E-11	5.4E-11	4.7E-11
Brain	5.1E-11	3.5E-12	1.7E-12	1.0E-12	6.0E-13	6.5E-13
Breast	5.4E-11	6.3E-12	3.1E-12	1.8E-12	9.3E-13	8.9E-13
GI-Tract						
St Wall	1.3E-08	7.2E-09	3.5E-09	2.0E-09	1.3E-09	1.1E-09
SI Wall	3.0E-08	1.9E-08	9.6E-09	5.7E-09	3.2E-09	2.6E-09
ULI Wall	1.8E-07	1.2E-07	5.8E-08	3.4E-08	1.9E-08	1.5E-08
LLI Wall	4.9E-07	3.2E-07	1.6E-07	9.5E-08	5.4E-08	4.3E-08
Kidneys	8.2E-10	6.8E-11	3.8E-11	2.5E-11	1.7E-11	1.9E-11
Liver	3.6E-10	4.2E-11	2.2E-11	1.3E-11	7.8E-12	7.9E-12
Lungs	5.7E-11	7.8E-12	3.6E-12	2.1E-12	1.2E-12	1.0E-12
Muscle	7.3E-11	2.0E-11	1.1E-11	7.1E-12	4.6E-12	3.9E-12
Ovaries	2.8E-10	1.7E-10	9.8E-11	6.6E-11	4.4E-11	3.3E-11
Pancreas	7.9E-11	2.4E-11	1.3E-11	8.1E-12	4.8E-12	3.8E-12
Red Marrow	4.8E-09	2.9E-10	1.4E-10	7.6E-11	4.5E-11	2.9E-11
Skin	5.9E-11	9.3E-12	4.9E-12	3.0E-12	1.9E-12	1.7E-12
Spleen	7.2E-11	1.9E-11	1.0E-11	6.4E-12	3.7E-12	3.1E-12
Testes	7.2E-11	2.5E-11	1.5E-11	1.1E-11	7.8E-12	4.6E-12
Thymus	5.3E-11	5.2E-12	2.6E-12	1.4E-12	8.4E-13	8.2E-13
Thyroid	5.2E-11	4.3E-12	2.0E-12	1.2E-12	6.5E-13	6.8E-13
Uterus	1.5E-10	7.7E-11	4.3E-11	2.8E-11	1.8E-11	1.4E-11
Remainder	8.3E-10	4.7E-10	2.5E-10	1.3E-10	7.2E-11	5.7E-11
Effective Dose	4.0E-08	2.5E-08	1.3E-08	7.5E-09	4.2E-09	3.4E-09

GI-Tract Gastrointestinal Tract
St Stomach
SI Small Intestine
ULI Upper Large Intestine
LLI Lower Large Intestine

* In the biokinetic model for Th parameter values for the adult apply to ages > 25 y. For radioisotopes of this element the dose coefficients for the adult are based on the 50-y integrated doses following an acute intake at age 25 y.

References

Boecker, B. B., Thomas, R. G. and Scott, J. K. (1963). Thorium distribution and excretion studies. II. General patterns following inhalation and the effect of the size of the inhaled dose. *Health Phys.* **9**, 165–176.

Dang, H. S. and Sunta, C. M. (1990). Gastrointestinal absorption factor (f_1) for Th from diet. *Health Phys.* **58**, 220–221.

Dang, H. S., Jaiswal, D. D., Murthy, K. B. S., Sharma, R. C., Nambiar, P. P. V. J. and Sunta, C. M. (1992). Relevance of ICRP metabolic model of Th in bio-assay monitoring. *J. Radioanalytical Nucl. Chem. Articles* **156**, 55–64.

Hamilton, J. G. (1947). The metabolism of the fission products and the heaviest elements. *Radiology* **49**, 325–343.

Hewson, G. S. and Fardy, J. J. (1993). Thorium metabolism and bioassay of mineral sands workers. *Health Phys.* **64**, 147–156.

Hunt, G. J., Leonard, D. R. P. and Lovett, M. B. (1986). Transfer of environmental plutonium and americium across the human gut. *Science Total Environ.* **53**, 89–109.

Hunt, G. J., Leonard, D. R. P. and Lovett, M. B. (1990). Transfer of environmental plutonium and americium across the human gut: A second study. *Science Total Environ.* **90**, 273–282.

Ibrahim, S. A., Wrenn, M. E., Singh, N. P., Cohen, N. and Saccomanno, G. (1983). Thorium concentration in human tissues from two US populations. *Health Phys.* **44** (Suppl. 1), 213–220.

ICRP (1975). *Report of the Task Group on Reference Man.* ICRP Publication 23, Pergamon Press, Oxford.

ICRP (1979). *Limits for Intakes of Radionuclides by Workers.* ICRP Publication 30, Part 1. *Annals of the ICRP* **2**(3/4), Pergamon Press, Oxford.

ICRP (1986). *The Metabolism of Plutonium and Related Elements.* ICRP Publication 48. *Annals of the ICRP* **16**(2/3), Pergamon Press, Oxford.

ICRP (1989). *Age-dependent Doses to Members of the Public from Intake of Radionuclides.* ICRP Publication 56, Part 1. *Annals of the ICRP* **20**(2), Pergamon Press, Oxford.

ICRP (1993). *Age-dependent Doses to Members of the Public from Intake of Radionuclides.* ICRP Publication 67, Part 2 Ingestion Dose Coefficient. *Annals of the ICRP* **23**(3/4), Elsevier Science Ltd, Oxford.

Jee, W. S. (1964). A critical survey of the analysis of microscopic distribution of some bone-seeking radionuclides and assessment of absorbed dose. In: *Assessment of Radioactivity in Man,* Vol. II (Vienna: IAEA) 369–393.

Jee, W. S. S., Arnold, J. S., Cochran, T. H., Twente, J. A. and Mical, R. S. (1962). Relationship of microdistribution of alpha particles to damage. In: *Some Aspects of Internal Irradiation,* pp. 27–45 (Dougherty, T. F., Jee, W. S. S., Mays, C. W. and Stover, B. J. eds), Pergamon Press, Oxford.

Johnson, J. R. and Lamothe, E. S. (1989). A review of the dietary uptake of Th. *Health Phys.* **56**, 165–168.

Larsen, R. P., Oldham, R. D., Bhattacharyya, M. H. and Moretti, E. S. (1984). Gastrointestinal absorption and distribution of thorium in the mouse. In: *Environmental Research Division Annual Report,* 65–68. July 1982–June 1983; Argonne National Laboratory, ANL-83-100-Pt. 2.

Maletskos, C. J., Keane, A. T., Telles, N. C. and Evans, R. D. (1966). The metabolism of intravenously administered radium and thorium in human beings and the relative absorption from the human gastrointestinal tract of radium and thorium in simulated radium dial paints. In: *Radium and Mesothorium Poisoning and Dosimetry and Instrumentation Techniques in Applied Radioactivity (MIT-952-3),* 202–317. Massachusetts Institute of Technology, Cambridge, Massachusetts.

Maletskos, C. J., Keane, A. T., Telles, N. C. and Evans, R. D. (1969). Retention and absorption of ^{224}Ra and ^{234}Th and some dosimetric considerations of ^{224}Ra in human beings. In: *Delayed Effects of Bone-seeking Radionuclides,* pp. 29–49 (Mays, C. W., Jee, W. S. S., Lloyd, R. D., Stover, B. J., Dougherty, J. H. and Taylor, G. N. eds). UT: University of Utah Press, Salt Lake City.

McInroy, J. F., Kathren, R. L. and Swint, M. J. (1990). Distribution of plutonium and americium in whole bodies donated to the United States Transuranium Registry. *Radiat. Prot. Dosim.* **26**, 151–158.

Miller, S. C., Bruenger, F. W. and Williams, F. W. (1989). Influence of age at exposure on concentrations of Pu-239 in beagle gonads. *Health Phys.* **56**, 485–491.

NEA/OECD (1988). *Committee on Radiation Protection and Public Health.* Report of an Expert Group on Gut Transfer Factors. NEA/OECD, Paris.

Newton, D., Rundo, J. and Eakins, J. D. (1981). Long-term retention of ^{228}Th following accidental intake. *Health Phys.* **40**, 291–298.

Pavlovskaya, N. A., Provotorov, A. V. and Makeeva, L. G. (1971). Absorption of thorium from the gastrointestinal tract into blood of rats and its buildup in organs and tissues. *Gig. Sanit.* **36** No. 5, 47–50.

Popplewell, D. S., Harrison, J. D. and Ham, G. J. (1991). Gastrointestinal absorption of neptunium and uranium in humans. *Health Phys.* **60**, 797–805.

Popplewell, D. S., Ham, G. J., McCarthy, W. and Lands, C. (1994). Transfer of plutonium across the human gut and its urinary excretion. In: *Intakes of Radionuclides.* CEC/US-DOE/NRPB/IPSN Workshop, Bath, September 1993.

Rundo, J. (1964). Two cases of chronic occupational exposure to radioactive materials. In: *Assessment of Radioactivity in Man,* Vol. II (Vienna: IAEA) pp. 291–306.

Scott, J. K., Neuman, W. F. and Bonner, J. F. (1952). The distribution and excretion of thorium sulphate. *J. Pharmacol. Exp. Ther.* **106**, 286–290.

Singh, N. P., Wrenn, M. E. and Ibrahim, S. A. (1983). Plutonium concentration in human tissues: Comparison to thorium. *Health Phys.* **44** (Suppl. 1), 469–476.

Singh, N. P., Zimmerman, C. J., Taylor, G. N. and Wrenn, M. E. (1988). The beagle: An appropriate experimental animal for extrapolating the organ distribution pattern of Th in humans. *Health Phys.* **54**, 293–299.

Stover, B. J., Atherton, D. R., Keller, N. and Buster, D. S. (1960). Metabolism of the Th-228 decay series in adult beagle dogs. *Radiat. Res.* **12**, 657–671.

Sullivan, M. F. (1980). Absorption of actinide elements from the gastrointestinal tract of rats, guinea pigs, pigs and dogs. *Health Phys.* **38**, 159–171.

Sullivan, M. F., Miller, B. M. and Ryan, J. L. (1983). Absorption of thorium and protactinium from the gastrointestinal tract in adult mice and rats and neonatal rats. *Health Phys.* **44**, 425–428.

Sunta, C. M., Dang, H. S., Jaiswal, D. D. and Soman, S. D. (1990). Thorium in human blood serum, clot, and urine, comparison with ICRP excretion model. *J. Radioanalytical Nucl. Chem. Articles* **138**, 139–144.

Thomas, R. G., Lie, R. and Scott, J. K. (1963). Thorium distribution and excretion studies. I. Patterns following parenteral administration. *Health Phys.* **9**, 153–163.

Traikovich, M. (1970). Absorption, distribution and excretion of certain soluble compounds of natural thorium. In: *Toxicology of Radioactive Substances,* Vol. 4: Thorium-232 and uranium-238 (Letavet, A. A. and Kurlyandskaya, E. B. eds). English translation (Dolphin, G. W. ed), Pergamon Press, Oxford.

Wrenn, M. E., Singh, N. P., Cohen, N., Ibrahim, S. A. and Saccomanno, G. (1990). *Thorium in Human Tissues,* New York University Medical Center, NUREG/CR-1227.

5. URANIUM

(123) Although this report is concerned with the calculation of dose coefficients it should be noted that intakes of the more transportable uranium compounds are limited by considerations of chemical toxicity, rather than radiation dose (ICRP, 1988).

Uptake to blood

(124) *Adults.* Data on the absorption of uranium have been recently reviewed by Wrenn *et al.* (1985), Harrison (1991) and Leggett and Harrison (1995).

(125) In the first controlled human study involving more than one subject, Hursh *et al.* (1969) administered uranyl nitrate to four hospital patients. The data obtained were taken to suggest fractional absorption in the range 0.005–0.05. Leggett and Harrison (1995) have interpreted the data as suggesting absorption of 0.004, 0.01, 0.02 and 0.06 for the four subjects. Wrenn *et al.* (1989) estimated absorption in 12 normal healthy adult volunteers given drinking water high in uranium. On the basis that 40–60% of absorbed uranium was excreted in the urine in the first 3 days after ingestion, rather than the authors assumption of 79%, Leggett and Harrison (1995) concluded that mean absorption was 0.006–0.015, maximum absorption was in the range 0.02–0.04, and absorption was less than 0.0025 in at least five subjects. Recently, Harduin *et al.* (1994) reported results for the absorption of uranium from drinking water either administered on 1 d or over 15 d. The data for acute administration suggested absorption of 0.005–0.05 with an average value of 0.015–0.02. The data for 15-d administration suggested absorption of 0.003–0.02 and average absorption of 0.01–0.015.

(126) A number of dietary balance studies have indicated mean fractional absorption values in the range 0.004–0.04 (Larsen and Orlandini, 1984; Spencer *et al.*, 1990; Wrenn *et al.*, 1989; Leggett and Harrison, 1995). Larsen and Orlandini (1984) measured daily excretion of uranium in two subjects whose drinking water had a high ^{234}U to ^{238}U activity ratio. Combining estimated intake and urinary excretion data for the two isotopes gives a range in absorption of 0.004–0.007. Spencer *et al.* (1990) determined uranium intake and excretion in four adult male subjects in a metabolism ward. The data were interpreted as suggesting that absorption of uranium from food was negligible while absorption from water was about 0.05. Their conclusion has been questioned, however (Leggett and Harrison, 1995). On the assumption that all ingested uranium is equally available for absorption, the data indicate mean absorption of 0.015 (range of 0.005–0.026).

(127) Measurements of uranium absorption have been made in rats, hamsters, rabbits, dogs and baboons (reviewed by Wrenn *et al.*, 1985; Harrison, 1991; Leggett and Harrison, 1995). A number of studies have shown that absorption is substantially greater in fasted than fed animals. For example, Bhattacharyya *et al.* (1989) found that uptake was increased by an order of magnitude in mice and baboons deprived of food for 24 h prior to uranium administration. The values obtained for absorption in mice were 7×10^{-4} in fed animals and 0.008 after fasting, with corresponding values for baboons of 0.005 and 0.045. Animal studies provide information on the relative uptake of uranium ingested in different chemical forms. Absorption generally decreases with decreasing solubility of the compound, being greatest for uranium ingested as $UO_2(NO_3)_2\ 6H_2O$,

UO_2F_2 or UO_2F_2, about half as great for UO_4 or UO_3 and one to two orders of magnitude lower for UCl_4, U_3O_8, UO_2 and UF_4.

(128) *ICRP Publication 30* (ICRP, 1979) adopted an f_1 value of 5×10^{-2} for inorganic forms of uranium relying mainly on the human data reported by Hursh *et al.* (1969). On the basis of more recent data, as reviewed by Wrenn *et al.* (1985), Harrison (1991) and Leggett and Harrison (1995), an f_1 value of 2×10^{-2} appears to be a more realistic value for dietary forms of uranium and is adopted here.

(129) *Children.* Limited data on the absorption of uranium in children from 5 y of age suggest that uptake does not vary substantially with age (Svatkina and Novikov, 1975; Limson-Zamora *et al.*, 1992; Leggett and Harrison, 1995). However, estimates were based on the assumption that subjects are in uranium balance and hence could underestimate uptake in rapidly growing children who may be expected to show net retention of uranium.

(130) Increased absorption of uranium has been demonstrated in neonatal rats and pigs (Sullivan, 1980; Sullivan and Gorham, 1982). Absorption in 2-d-old rats given uranium nitrate was estimated as about 0.01–0.07, two orders of magnitude greater than for adults. In pigs given uranium nitrate on the first day after birth, retention in the skeleton 7 d later accounted for about 30% of the administered uranium. Particularly high absorption in pigs is consistent with the high permeability of the gut in this species to the absorption of intact immunoglobulins in milk, occurring during the first 1–2 d of life (Brambell, 1970).

(131) On the basis of these data and in the absence of quantitative information on human infants, the general approach of the NEA expert group (NEA/OECD, 1988) is adopted here (see Paragraph (8) of the Introduction) to give an f_1 value for the 3-mo-old infant of 4×10^{-2}. For children of 1 y and older, the f_1 value for the adult of 2×10^{-2} is adopted here.

Distribution and retention

(132) Although uranium exists in several oxidation states the hexavalent form is the one commonly encountered in environmental exposures. Consequently, attention has been restricted to data on hexavalent uranium, insofar as the form of uranium was known or reported.

(133) Uranium entering the blood is rapidly taken up by tissues or excreted in urine. Typically, 25% of intravenously injected uranium administered as the nitrate remained in blood of human subjects after 5 min, 5% after 5 h, 1% after 20 h, and less than 0.5% after 100 h, but inter-subject variation was high (Bassett *et al.*, 1948; Bernard *et al.*, 1956; Struxness *et al.*, 1956; Bernard and Struxness, 1957; Luessenhop *et al.*, 1958). Data on laboratory animals indicate that a substantial fraction of plasma uranium is associated with the ultrafilterable low molecular weight fraction and the remainder is weakly associated with transferrin and other plasma proteins (Stevens *et al.*, 1980; Cooper *et al.*, 1982; Durbin, 1984). Limited measurements on blood from human donors exposed only to environmental uranium indicate that most uranium in the blood is associated with red blood cells (Lucas and Marcun, 1970; Fisenne and Welford, 1985). Data on baboons indicate that 50% or more of the uranium in blood is associated with the red blood cells during the period 10–1000 h after injection (Lipsztein, 1981). These data have been interpreted as indicating that about 0.7% of uranium leaving plasma attaches to red blood cells and is returned to plasma with a half-time slightly greater than 1 d (Lipsztein, 1981).

(134) In experimental studies on humans, typically two-thirds of uranium intravenously injected as the nitrate was excreted in urine in the first 24 h and roughly a further 10% over the next 5 d (Bassett *et al.*, 1948; Bernard *et al.*, 1957; Bernard and Struxness, 1957; Struxness *et al.*, 1956; Luessenhop *et al.*, 1958; Terepka *et al.*, 1964). Similar results were obtained for baboons (Lipsztein, 1981). In dogs, urinary excretion accounted for 22–58% of uranium injected as the nitrate after 24 h and 80–90% of injected uranium after 2–3 weeks (Stevens *et al.*, 1980; Morrow *et al.*, 1982). Data on man and laboratory animals indicate that most of the remaining uranium is excreted over a period of a few months, but a few per cent of the amount injected may be retained for a period of years (Bernard *et al.*, 1956; Struxness *et al.*, 1956; Luessenhop *et al.*, 1958; Stevens *et al.*, 1980; Sontag, 1984). Based on the results of a study involving intravenous injection of uranyl nitrate in three control patients and seven patients with various bone disorders, Terepka and co-workers (see Hursh and Spoor, 1973) concluded that in cases where there was an increase in exchangeable bone (Paget's disease and osteomalacia) urinary excretion of uranium decreases and deposition in bone probably increases.

(135) During the first few days after intravenous injection of uranyl initiate into patients who were in the latter stages of terminal diseases of the central nervous system (the "Boston subjects"), faecal excretion accounted for less than 1% of total excretion (Bernard, 1956). Similar results were obtained for baboons (Lipsztein, 1981). In beagles, an estimated 2–5% of injected uranium is excreted in faeces during the first 2 weeks (Stevens *et al.*, 1980; Morrow *et al.*, 1982).

(136) Data on laboratory animals indicate that a substantial portion of uranium leaving blood may initially distribute throughout the soft tissues, but by a few days after absorption or injection into the blood, most of the systemic content is found in the kidneys and skeleton (Stevens *et al.*, 1980; Lipsztein, 1981; Morrow *et al.*, 1982; Sontag, 1984; Bhattacharyya *et al.*, 1989). For example, the kidneys and skeleton of beagles injected with uranium contained nearly half of the systemic burden after 1 d and roughly 90% at 2–6 d (Morrow *et al.*, 1982). Based in part on considerations of material balance, it is estimated that the skeleton and kidneys of a baboon injected with uranium contained 80–90% of the systemic burden at 4 d (Lipsztein, 1981; Durbin, 1984).

(137) Measurements of the systemic distribution of uranium were made at autopsy in the "Boston subjects" (Bernard *et al.*, 1956; Bernard and Struxness, 1957; Struxness *et al.*, 1956; Luessenhop *et al.*, 1958). The skeleton, kidneys, and other soft tissues of a subject dying 2.5 d after injection contained about 10%, 14%, and 6%, respectively, of the administered amount. In a subject dying 18 d after injection, the bones, kidneys, and other soft tissue contained about 4–13%, 6%, and 4%, respectively, of the administered amount. Data for a subject dying at 566 d after injection indicated that the contents of the skeleton, kidneys, and other soft tissues had declined to about 1.4%, 0.3%, and 0.3%, respectively, of the administered amount.

(138) A substantial portion of uranium filtered by the kidneys is temporarily retained in the renal tubules before passing to the urinary bladder contents. Morrow *et al.* (1982) estimated that the kidneys of beagle dogs contained 44% of absorbed uranium at 6 h after inhalation of UO_2F_2 and 16% of the absorbed amount at 24 h. At 1–3 d after inhalation or injection of soluble forms of uranium, the kidneys of humans, dogs, and rats contained 12–25% of the amount reaching blood (Bernard and Struxness, 1957; Muir *et al.*, 1960; Jones, 1966; Stevens *et al.*, 1980; Morrow *et al.*, 1982). About 20–40% of injected uranium was found in the kidneys of mice at 1 d (Kisieleski *et al.*, 1952). Available data indicate that retention of uranium in the kidneys as a function of time after

administration of moderate to high amounts is broadly similar in humans, dogs, and rats. Durbin (1984) reviewed data on retention of uranium in the kidneys of humans, beagles, rats, and mice and concluded that 92–95% of the renal content remaining at 1 d is lost with a half-time of 2–6 d and the remainder is lost with a half-time of 30–340 d. Retention of uranium by the kidneys appears to increase with the mass absorbed to blood, at least at relatively high masses of uranium (Bernard, 1956; Luessenhop, 1958; Durbin, 1984; Leggett, 1989). This complicates the development of a biokinetic model for uranium, because relatively high masses of uranium have been administered in most experimental studies of the biokinetics of uranium, including the "Boston study". Moreover, some of the above studies employing uranium isotopes with relatively high specific activity have involved high radiation doses to the kidneys and bone surfaces. High mass or radiation damage may have led to kinetics different from that occurring at low mass and low dose (Stevens *et al.*, 1980).

(139) Uptake and retention of uranium by the liver can be estimated from the reasonably consistent data developed for the "Boston subjects", dogs, and baboons. These data indicate that the liver accumulates about 1.5–2% of injected uranium, that most of this is removed over the first few weeks after injection, and that roughly 0.1% of injected uranium is retained more tenaciously, being lost over a period of months or years.

(140) There is a fairly large body of autopsy data on persons chronically exposed to uranium, either environmentally or occupationally (Donoghue, 1972; Campbell, 1975; Roberts *et al.*, 1977; Igarashi *et al.*, 1985; Fisenne and Welford, 1986; Singh *et al.*, 1986, 1987; Kathren *et al.*, 1989). Because of the different and generally unknown exposure patterns experienced by the subjects, it is useful to normalise the data to the uranium content of liver. The liver was chosen for this purpose rather than the kidneys or skeleton because the uranium content of liver is generally measurable but low enough to avoid important mass or radiation effects, and because fewer sampling problems are associated with the liver than with the skeleton. The normalised data suggest that the skeleton and kidneys typically retain, respectively, about 30 and 0.5 times as much uranium as the liver. With regard to skeletal uranium, the measurements are usually confined to rib and vertebrae (which contain a high proportion of trabecular bone), with only occasional measurements of femur and other bones, and hence may not be representative of the entire skeleton.

(141) Autopsy data on the "Boston subjects" indicate that the soft tissues other than liver and kidneys contain about 5% of injected uranium in the first few days after administration, 0.5–2.5% at 25–139 d, and 0.3% at 1.5 y. Following chronic exposure to uranium, the soft tissues other than the liver and kidneys may account for a non-trivial portion of the total-body content of uranium (Fisenne *et al.*, 1989; Gonzales and McInroy, 1991). For example, in two subjects not exposed occupationally to uranium, muscle and skin accounted for about 25% of systemic uranium, compared with about 70% in the skeleton (Gonzales and McInroy, 1991).

(142) In man and adult animals, 4–25% of absorbed or injected uranium is found in the skeleton in the first week after administration. Autopsy measurements on persons previously exposed occupationally to uranium as well as persons exposed only to environmental uranium suggest that bone is the primary site of long-term retention although accumulating, at most, a few per cent of activity reaching the systemic circulation.

(143) The skeletal behaviour of uranium is in some ways qualitatively similar to that of the alkaline earths. It is known that UO_2^{2+} exchanges with Ca^{2+} on the surfaces of

bone mineral crystals although it does not participate in crystal formation or enter existing crystals. The early distribution of uranium among different parts of the skeleton is similar to that of calcium. Uranium initially deposits on all bone surfaces but is most highly concentrated in areas of growth. Perhaps depending on the microscopic structure of the bone of each species, uranium on bone surfaces may gradually diffuse into bone volume; such diffusion has been observed in dogs (Rowland and Farnham, 1968; Stevens *et al.*, 1980) but not in rats (Priest *et al.*, 1982) or mice (Kisieleski *et al.*, 1952). As is the case for calcium, a substantial portion of uranium deposited in bone is lost to plasma by processes that occur more rapidly than bone resorption. In the "Boston subjects" injected with uranium, an estimated 80–90% of the original skeletal deposition was lost from bone over the first 1.5 y.

(144) Despite numerous studies on the metabolic behaviour of uranium in laboratory animals and more limited studies in man, substantial uncertainties remain regarding long-term retention of this element in bone as well as soft tissues. In the biokinetic model for uranium applied in this document, parameter values describing long-term retention were chosen for consistency not only with the available experimental data but also with autopsy data on persons chronically exposed to uranium (either environmentally or occupationally). The reliability of these parameter values is limited by uncertainties in the level and pattern of uptake of uranium by the subjects. The present model yields somewhat higher estimates of long-term retention in bone, liver, and other soft tissues than does a recently developed model of Wrenn *et al.* (1994) in which retention is based on relatively short-term data. For purposes of this document, however, it is desirable to achieve maximum consistency with chronic exposure data, particularly environmental data.

Parameter values for adults

(145) The studies described above demonstrate that the behaviour of uranium in the body can be influenced by the mass concentration. In choosing model parameters it is important to select values that are appropriate for environmental concentrations of uranium. For workers exposed occupationally to high concentrations of uranium alternative parameter values may be needed.

(146) Because uranium tends to follow the qualitative behaviour of calcium to a large extent with regard to skeletal kinetics, the generic model structure for alkaline earth elements described in *ICRP Publication 67* (ICRP, 1993) is applied to uranium (see Fig. 5.1). Because some transfer rates in the biokinetic model for uranium are equated with bone formation rates which are expected to remain elevated during the early part of the third decade of life the dose coefficients for the adult are based on equivalent dose rates received over a 50 y period following acute intake at age 25 y.

(147) *Circulation:* Data on the early behaviour of uranium in the human circulation can be represented reasonably well by treating plasma as a uniformly mixed pool from which uranium is removed at a rate of 35 d^{-1} (one half the transfer rate used for radium in *ICRP Publication 67*) and postulating a soft tissue compartment, ST0, in relatively rapid exchange with plasma. Compartment ST0 serves two purposes in this model: to help maintain the proper amount of uranium in plasma, and to help depict an early build-up and decline of uranium in soft tissue other than liver and kidneys. Compartment ST0 is assumed to receive 30% of uranium leaving plasma; this is a default value used in the generic model when data are sparse and considerations such as balance of material do not prevail. The assumed removal half-time from ST0 to plasma is 2 h.

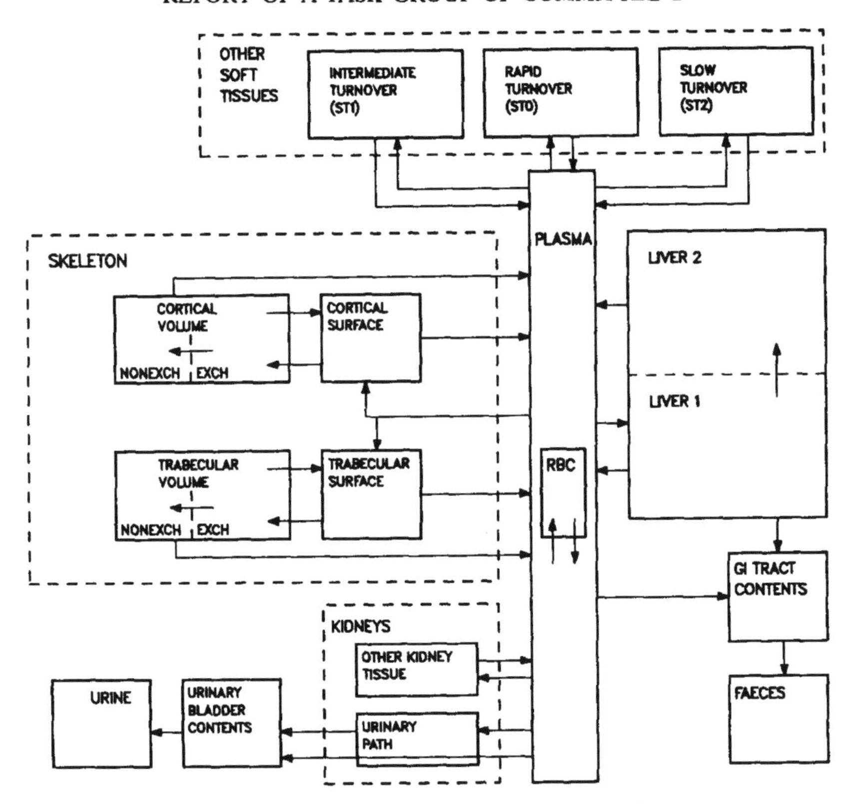

Fig. 5.1. Diagram of the biokinetic model for uranium.

(148) Transfer rates from plasma to the various compartments other than the rapid-turnover soft tissue compartment (ST0) are derived from the assumed total outflow rate from plasma (35 d^{-1}) and assumptions concerning the instantaneous partition of uranium that leaves the circulation. The term "leaving the circulation" refers to activity leaving plasma but not depositing in compartment ST0. Transfer rates from compartments other than plasma are generally based on assumed "removal half-times" which refer to the biological half-time that would be observed with no recycling to that compartment. This will generally differ from the apparent (or net, or externally viewed) half-time that may be estimated at any given time in the presence of recycling.

(149) Based on data for baboons (Lipsztein, 1981), it is assumed that 1% of uranium leaving the circulation (or 0.7% of uranium leaving plasma) deposits in the red blood cells. The removal half-time from red blood cells to plasma is assumed to be 2 d; this value was derived from the assumed level of transfer from plasma to red blood cells and the assumption that for chronic exposure roughly half of uranium in the blood is found in red blood cells.

(150) *Excretion:* Urinary excretion is assumed to arise from: (1) uranium moving directly from plasma to the urinary bladder contents, accounting for 63% of uranium leaving the circulation and (2) uranium moving to the urinary bladder contents after temporary residence in the renal tubules, accounting for 12% of uranium leaving the circulation. Uranium is assumed to be removed from the renal tubules to the urinary bladder with a half-time of 7 d. After parameter values for the renal tubules (urinary

path in Fig. 5.1) and liver were selected, parameter values describing uptake and removal by other kidney tissue were set for consistency with the relative amounts of uranium measured in liver and kidneys of environmentally exposed persons. Other kidney tissue is assumed to receive 0.05% of uranium leaving circulation. Uranium is assumed to be removed from other kidney tissue to plasma with a half-time of 5 y.

(151) It is assumed that 0.5% of uranium leaving the circulation enters the upper large intestine and is subsequently excreted in the faeces.

(152) *Liver:* The liver is assumed to consist of two compartments, Liver 1 and Liver 2. Liver 1 is assumed to receive 1.5% of uranium leaving the circulation. The removal half-time from Liver 1 is 7 d; 93% of the uranium leaving Liver 1 is assumed to return to plasma and 7% is assumed to move to Liver 2, from which it is returned to plasma with a half-time of 10 y.

(153) *Other soft tissues:* In addition to compartment ST0, two other soft tissue compartments are used to describe uptake and retention by soft tissue other than liver and kidneys. These are referred to as intermediate (ST1) and slow (ST2) turnover compartments of other soft tissue. Parameter values for ST1 and ST2 are based primarily on the data for the Boston subjects, but chronic exposure data are also considered. In particular, a very long removal half-time from compartment ST2 is assumed, based on recent data indicating that as much as one-quarter of systemic uranium could be retained in soft tissue other than liver and kidneys of chronically exposed persons (Gonzales and McInroy, 1991). Compartment ST1 is assumed to receive 6.65% of uranium leaving the circulation, which is the amount left over after other deposition fractions in the model had been assigned. The removal half-time from ST1 to plasma is assumed to be 20 d. Compartment ST2 is assumed to receive 0.3% of uranium leaving the circulation. The removal half-time from ST2 to plasma is assumed to be 100 y.

(154) *Skeleton:* Parameter values for skeleton are based on uranium intravenous injection data for adult man and laboratory animals, data on humans chronically exposed (normalised to liver), and analogy with radium. It is assumed that 15% of uranium leaving the circulation deposits on bone surfaces. By analogy with the alkaline earths (ICRP, 1993), the ratio of the amount deposited on trabecular surfaces to that deposited on cortical surfaces is assumed to be 1.25 in the mature skeleton. The removal half-time from bone surfaces is assumed to be 5 d. It is assumed that one-half of uranium leaving bone surfaces returns to plasma and one-half goes to an exchangeable bone volume compartment. Uranium is assumed to leave the exchangeable bone volume compartment with a half-time of 30 d; this value was derived for radium (ICRP, 1993) and is consistent with the limited data on loss of uranium from bone. Based on the small amount of uranium retained in the skeleton over a long term (as indicated by the human chronic exposure data) 75% of uranium leaving exchangeable bone is assumed to return to the bone surface compartment and 25% is assigned to non-exchangeable bone. Removal from non-exchangeable bone to plasma is at the age-dependent rate of bone turnover, which is different for cortical and trabecular bone (ICRP, 1989, 1993). As described in *ICRP Publication 67*, the transfer rates for the adult apply to ages ≥ 25 y because bone formation rates are expected to remain elevated during the early part of the third decade of life.

Parameter values for children

(155) There is little direct information on the kinetics of uranium in children. Measurements of uranium in bones of persons exposed only to environmental uranium

indicate that the concentration of uranium is higher in children than in adults (Broadway and Strong, 1983; Lianqing and Guiyun, 1990). Data on adult humans injected with uranium for experimental purposes indicate that deposition of uranium is higher in metabolically active skeletons than in relatively inactive skeletons (Terepka *et al.*, 1964) and hence is expected to be higher in immature than mature human skeletons, as has been demonstrated for rats (Neuman *et al.*, 1948).

(156) Age-specific deposition of uranium in the skeleton is assumed to be proportional to the deposition of the alkaline earth elements, and the rate of removal from deep bone is assumed to be the same as the age-specific bone turnover rate. As in the models for the alkaline earths as described in *ICRP Publication* 67 (ICRP, 1993), deposition of uranium in soft tissues and excreta is assumed to be reduced in children due to elevated uptake by bone. The method of derivation of deposition fractions for children is the same as described for radium (ICRP, 1993).

(157) As an example, consider the extension of parameter values from mature adult to age 10 y. At age 10 y, the rate of accretion of calcium by the skeleton is estimated as 0.56 g d^{-1}, compared with 0.25 g d^{-1} for the average mature adult (Leggett, 1992). The skeletal deposition fraction at age 10 y is estimated as the ratio of the accretion rates (0.56/0.25) times the value for mature adults (0.15), i.e. $0.15 \times 0.56/0.25 = 0.336$. Thus, for age 10 y, it is assumed that 33.6% of uranium leaving the circulation deposits on bone surfaces. The division between trabecular and cortical bone surfaces at any age depends on the age-specific calcium addition rate for that bone type. The trabecular to cortical deposition ratio increases gradually from 0.25 in infants to 1.25 in mature adults and is estimated as 0.34 at age 10 y (Leggett, 1992). Thus, for age 10 y, cortical bone is assumed to receive 33.6%/1.34 = 25.1% of uranium leaving the circulation and trabecular bone is assumed to receive 33.6 − 25.1% = 8.5%. To balance the elevated uptake of uranium by immature bone at age 10 y, deposition in ST1, ST2, liver, kidneys, urinary bladder, and gastrointestinal tract contents are decreased by the factor $(1.0 - 0.336)/(1.0 - 0.15) = 0.781$.

(158) Bone turnover rates assumed in the model are higher in children than in adults (Leggett, 1992), see Table 5.-1. In the absence of age-dependent information removal half-times in soft tissues are assumed to be the same for infants and children as for adults.

Behaviour of decay products

(159) The treatment of decay products is similar to that described in this document for thorium decay products. For treatment of thorium, actinium or protactinium produced *in vivo*, compartments for testes, ovaries, trabecular bone marrow, and cortical bone marrow must be added to the model structure shown in Fig. 5.1. Thorium or radium produced in red blood cells is assumed to move to blood plasma with a half-time of 1 d.

(160) Figures 5.2 and 5.3 give estimated uranium contents of bone and kidneys in infants, 10-y-old children, and adults as a function of time after injection, based on the parameter values given in Table 5.-1.

Dose coefficients

(161) Dose coefficients derived from the biokinetic data summarised in Table 5.-1 are given in Tables 5.-2 to 5.-7.

Table 5.-1. Age-specific transfer rates (d^{-1}) for uranium model

	Age					
	3 mo	1 y	5 y	10 y	15 y	Adult
Plasma to STO	1.050E+01	1.050E+01	1.050E+01	1.050E+01	1.050E+01	1.050E+01
Plasma to RBC	1.590E-01	2.100E-01	2.190E-01	1.910E-01	1.600E-01	2.450E-01
Plasma to Urinary bladder	9.990E+00	1.326E+01	1.380E+01	1.206E+01	1.010E+01	1.543E+01
Plasma to Urinary path	1.900E+00	2.520E+00	2.630E+00	2.300E+00	1.920E+00	2.940E+00
Plasma to Other kidney tissue	7.900E-03	1.050E-02	1.100E-02	9.600E-03	8.000E-03	1.220E-02
Plasma to ULI contents	7.900E-02	1.050E-01	1.100E-01	9.600E-02	8.000E-02	1.220E-01
Plasma to Liver 1	2.380E-01	3.160E-01	3.290E-01	2.870E-01	2.400E-01	3.670E-01
Plasma to ST1	1.050E+00	1.400E+00	1.460E+00	1.270E+00	1.070E+00	1.630E+00
Plasma to ST2	4.760E-02	6.310E-02	6.570E-02	5.740E-02	4.810E-02	7.350E-02
Plasma to Trabecular surfaces	2.200E+00	1.320E+00	1.310E+00	2.070E+00	3.030E+00	2.040E+00
Plasma to Cortical surfaces	8.820E+00	5.290E+00	4.570E+00	6.160E+00	7.840E+00	1.630E+00
STO to Plasma	8.320E+00	8.320E+00	8.320E+00	8.320E+00	8.320E+00	8.320E+00
RBC to Plasma	3.470E-01	3.470E-01	3.470E-01	3.470E-01	3.470E-01	3.470E-01
Other kidney tissue to Plasma	3.800E-04	3.800E-04	3.800E-04	3.800E-04	3.800E-04	3.800E-04
Liver 1 to Plasma	9.200E-02	9.200E-02	9.200E-02	9.200E-02	9.200E-02	9.200E-02
Liver 2 to Plasma	1.900E-04	1.900E-04	1.900E-04	1.900E-04	1.900E-04	1.900E-04
ST1 to Plasma	3.470E-02	3.470E-02	3.470E-02	3.470E-02	3.470E-02	3.470E-02
ST2 to Plasma	1.900E-05	1.900E-05	1.900E-05	1.900E-05	1.900E-05	1.900E-05
Bone surfaces to Plasma	6.930E-02	6.930E-02	6.930E-02	6.930E-02	6.930E-02	6.930E-02
Nonexch trab. vol. to Plasma	8.220E-03	2.880E-03	1.810E-03	1.320E-03	9.590E-04	4.930E-04
Nonexch cort. vol. to Plasma	8.220E-03	2.880E-03	1.530E-03	9.040E-04	5.210E-04	8.210E-05
Urinary path to Urinary bladder	9.900E-02	9.900E-02	9.900E-02	9.900E-02	9.900E-02	9.900E-02
Liver 1 to Liver 2	6.930E-03	6.930E-03	6.930E-03	6.930E-03	6.930E-03	6.930E-03
Bone surfaces to Exch. volume	6.930E-02	6.930E-02	6.930E-02	6.930E-02	6.930E-02	6.930E-02
Exch. bone vol. to Bone surfaces	1.730E-02	1.730E-02	1.730E-02	1.730E-02	1.730E-02	1.730E-02
Exch. bone vol. to Nonexch vol.	5.780E-03	5.780E-03	5.780E-03	5.780E-03	5.780E-03	5.780E-03
f_1	4.000E-02	2.000E-02	2.000E-02	2.000E-02	2.000E-02	2.000E-02

Parameters are given to sufficient precision for calculational purposes. This may be more precise than the biological data would support.

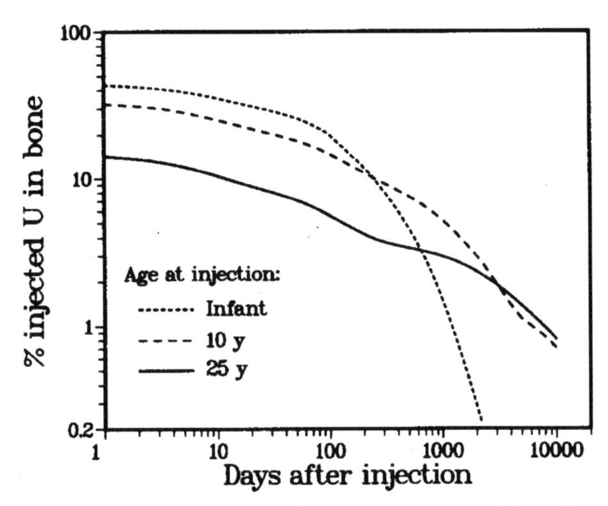

Fig. 5.2. Model predictions of the uranium content of bone as a function of time after injection into blood for different ages.

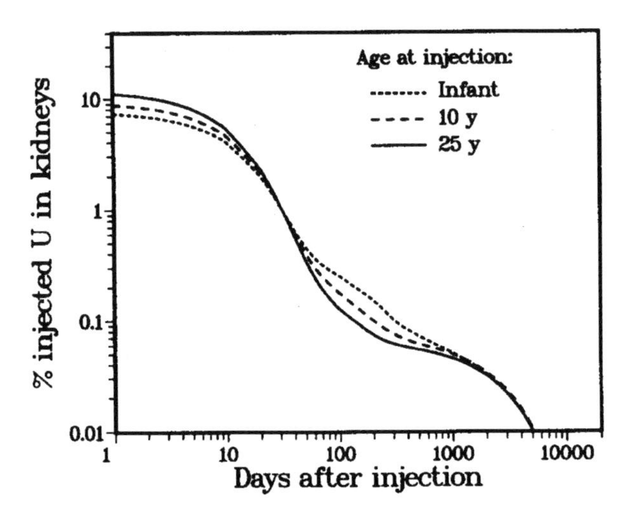

Fig. 5.3. Model predictions of the uranium contents of the kidneys as a function of time after injection into blood for different ages.

Table 5.-2.

Ingestion Dose Coefficients: Committed Equivalent and Effective Doses per Unit Intake (Sv/Bq) for U-232 (T1/2 = 72 y)*

Age at intake	3 Months	1 Year	5 Years	10 Years	15 Years	Adult
Adrenals	7.6E-07	3.3E-07	2.5E-07	2.1E-07	1.9E-07	1.7E-07
Bladder Wall	7.6E-07	3.3E-07	2.5E-07	2.1E-07	1.9E-07	1.7E-07
Bone Surfaces	5.8E-05	1.6E-05	1.2E-05	1.6E-05	2.3E-05	7.2E-06
Brain	7.6E-07	3.3E-07	2.5E-07	2.1E-07	1.9E-07	1.7E-07
Breast	7.6E-07	3.3E-07	2.5E-07	2.1E-07	1.9E-07	1.6E-07
GI-Tract						
St Wall	7.7E-07	3.3E-07	2.5E-07	2.1E-07	1.9E-07	1.7E-07
SI Wall	7.9E-07	3.5E-07	2.6E-07	2.2E-07	1.9E-07	1.7E-07
ULI Wall	1.0E-06	4.8E-07	3.3E-07	2.6E-07	2.2E-07	1.9E-07
LLI Wall	1.5E-06	7.7E-07	4.7E-07	3.5E-07	2.9E-07	2.3E-07
Kidneys	5.4E-06	2.0E-06	1.3E-06	1.0E-06	8.8E-07	6.6E-07
Liver	4.5E-06	1.7E-06	1.2E-06	9.7E-07	9.6E-07	6.9E-07
Lungs	7.6E-07	3.3E-07	2.5E-07	2.1E-07	1.9E-07	1.6E-07
Muscle	7.6E-07	3.3E-07	2.5E-07	2.1E-07	1.9E-07	1.6E-07
Ovaries	9.7E-07	3.9E-07	3.1E-07	2.7E-07	2.5E-07	1.8E-07
Pancreas	7.6E-07	3.3E-07	2.5E-07	2.1E-07	1.9E-07	1.7E-07
Red Marrow	8.2E-06	2.1E-06	1.4E-06	1.4E-06	1.6E-06	7.1E-07
Skin	7.6E-07	3.3E-07	2.5E-07	2.1E-07	1.9E-07	1.6E-07
Spleen	7.6E-07	3.3E-07	2.5E-07	2.1E-07	1.9E-07	1.7E-07
Testes	1.1E-06	4.2E-07	3.3E-07	3.1E-07	2.6E-07	1.8E-07
Thymus	7.6E-07	3.3E-07	2.5E-07	2.1E-07	1.9E-07	1.6E-07
Thyroid	7.6E-07	3.3E-07	2.5E-07	2.1E-07	1.9E-07	1.6E-07
Uterus	7.6E-07	3.3E-07	2.5E-07	2.1E-07	1.9E-07	1.6E-07
Remainder	8.4E-07	3.5E-07	2.7E-07	2.2E-07	2.0E-07	1.7E-07
Effective Dose	2.5E-06	8.2E-07	5.9E-07	5.8E-07	6.4E-07	3.4E-07

GI-Tract Gastrointestinal Tract
St Stomach
SI Small Intestine
ULI Upper Large Intestine
LLI Lower Large Intestine

* In the biokinetic model for U parameter values for the adult apply to ages > 25 y. For radioisotopes of this element the dose coefficients for the adult are based on the 50-y integrated doses following an acute intake at age 25 y.

Table 5.-3.

Ingestion Dose Coefficients: Committed Equivalent and Effective Doses per Unit Intake (Sv/Bq) for U-233 (T1/2 = 1.585E5 y)*

Age at intake	3 Months	1 Year	5 Years	10 Years	15 Years	Adult
Adrenals	1.3E-07	6.0E-08	4.5E-08	3.6E-08	3.0E-08	2.8E-08
Bladder Wall	1.3E-07	6.0E-08	4.5E-08	3.6E-08	3.0E-08	2.8E-08
Bone Surfaces	7.8E-06	1.8E-06	1.3E-06	1.5E-06	2.4E-06	8.0E-07
Brain	1.3E-07	6.0E-08	4.5E-08	3.6E-08	3.0E-08	2.8E-08
Breast	1.3E-07	6.0E-08	4.5E-08	3.6E-08	3.0E-08	2.8E-08
GI-Tract						
St Wall	1.5E-07	6.8E-08	4.9E-08	3.8E-08	3.2E-08	2.9E-08
SI Wall	1.7E-07	8.1E-08	5.6E-08	4.2E-08	3.4E-08	3.1E-08
ULI Wall	3.3E-07	1.9E-07	1.1E-07	7.3E-08	5.1E-08	4.5E-08
LLI Wall	6.8E-07	4.2E-07	2.3E-07	1.4E-07	9.1E-08	7.7E-08
Kidneys	2.8E-06	1.1E-06	6.4E-07	4.4E-07	3.3E-07	2.9E-07
Liver	5.9E-07	2.6E-07	1.9E-07	1.4E-07	1.1E-07	1.1E-07
Lungs	1.3E-07	6.0E-08	4.5E-08	3.6E-08	3.0E-08	2.8E-08
Muscle	1.3E-07	6.0E-08	4.5E-08	3.6E-08	3.0E-08	2.8E-08
Ovaries	1.3E-07	6.0E-08	4.5E-08	3.6E-08	3.0E-08	2.8E-08
Pancreas	1.3E-07	6.0E-08	4.5E-08	3.6E-08	3.0E-08	2.8E-08
Red Marrow	9.3E-07	2.1E-07	1.3E-07	1.3E-07	1.4E-07	8.2E-08
Skin	1.3E-07	6.0E-08	4.5E-08	3.6E-08	3.0E-08	2.8E-08
Spleen	1.3E-07	6.0E-08	4.5E-08	3.6E-08	3.0E-08	2.8E-08
Testes	1.3E-07	6.0E-08	4.5E-08	3.6E-08	3.0E-08	2.8E-08
Thymus	1.3E-07	6.0E-08	4.5E-08	3.6E-08	3.0E-08	2.8E-08
Thyroid	1.3E-07	6.0E-08	4.5E-08	3.6E-08	3.0E-08	2.8E-08
Uterus	1.3E-07	6.0E-08	4.5E-08	3.6E-08	3.0E-08	2.8E-08
Remainder	1.8E-07	7.8E-08	5.5E-08	4.1E-08	3.3E-08	3.1E-08
Effective Dose	3.7E-07	1.3E-07	8.9E-08	7.5E-08	7.5E-08	5.0E-08

GI-Tract Gastrointestinal Tract
St Stomach
SI Small Intestine
ULI Upper Large Intestine
LLI Lower Large Intestine

* In the biokinetic model for U parameter values for the adult apply to ages > 25 y. For radioisotopes of this element the dose coefficients for the adult are based on the 50-y integrated doses following an acute intake at age 25 y.

Table 5.-4.

Ingestion Dose Coefficients: Committed Equivalent and Effective Doses per Unit Intake (Sv/Bq) for U-234 (T1/2 = 2.445E5 y)*

Age at intake	3 Months	1 Year	5 Years	10 Years	15 Years	Adult
Adrenals	1.3E-07	5.9E-08	4.5E-08	3.5E-08	3.0E-08	2.8E-08
Bladder Wall	1.3E-07	6.0E-08	4.5E-08	3.5E-08	3.0E-08	2.8E-08
Bone Surfaces	7.7E-06	1.8E-06	1.3E-06	1.5E-06	2.3E-06	7.9E-07
Brain	1.3E-07	5.9E-08	4.5E-08	3.5E-08	3.0E-08	2.8E-08
Breast	1.3E-07	5.9E-08	4.5E-08	3.5E-08	3.0E-08	2.8E-08
GI-Tract						
St Wall	1.5E-07	6.7E-08	4.8E-08	3.7E-08	3.1E-08	2.9E-08
SI Wall	1.6E-07	8.0E-08	5.5E-08	4.1E-08	3.3E-08	3.0E-08
ULI Wall	3.2E-07	1.9E-07	1.1E-07	7.3E-08	5.0E-08	4.4E-08
LLI Wall	6.8E-07	4.2E-07	2.2E-07	1.4E-07	9.0E-08	7.6E-08
Kidneys	2.8E-06	1.0E-06	6.3E-07	4.4E-07	3.2E-07	2.9E-07
Liver	5.8E-07	2.6E-07	1.8E-07	1.4E-07	1.1E-07	1.1E-07
Lungs	1.3E-07	5.9E-08	4.5E-08	3.5E-08	3.0E-08	2.8E-08
Muscle	1.3E-07	5.9E-08	4.5E-08	3.5E-08	3.0E-08	2.8E-08
Ovaries	1.3E-07	6.0E-08	4.5E-08	3.5E-08	3.0E-08	2.8E-08
Pancreas	1.3E-07	5.9E-08	4.5E-08	3.5E-08	3.0E-08	2.8E-08
Red Marrow	9.1E-07	2.1E-07	1.3E-07	1.2E-07	1.4E-07	8.1E-08
Skin	1.3E-07	5.9E-08	4.5E-08	3.5E-08	3.0E-08	2.8E-08
Spleen	1.3E-07	5.9E-08	4.5E-08	3.5E-08	3.0E-08	2.8E-08
Testes	1.3E-07	5.9E-08	4.5E-08	3.5E-08	3.0E-08	2.8E-08
Thymus	1.3E-07	5.9E-08	4.5E-08	3.5E-08	3.0E-08	2.8E-08
Thyroid	1.3E-07	5.9E-08	4.5E-08	3.5E-08	3.0E-08	2.8E-08
Uterus	1.3E-07	5.9E-08	4.5E-08	3.5E-08	3.0E-08	2.8E-08
Remainder	1.8E-07	7.7E-08	5.4E-08	4.0E-08	3.3E-08	3.0E-08
Effective Dose	3.7E-07	1.3E-07	8.8E-08	7.4E-08	7.5E-08	5.0E-08

GI-Tract Gastrointestinal Tract
St Stomach
SI Small Intestine
ULI Upper Large Intestine
LLI Lower Large Intestine

* In the biokinetic model for U parameter values for the adult apply to ages > 25 y. For radioisotopes of this element the dose coefficients for the adult are based on the 50-y integrated doses following an acute intake at age 25 y.

Table 5.-5.

Ingestion Dose Coefficients: Committed Equivalent and Effective Doses per Unit Intake (Sv/Bq) for U-235 (T1/2 = 703.8E6 y)*

Age at intake	3 Months	1 Year	5 Years	10 Years	15 Years	Adult
Adrenals	1.2E-07	5.5E-08	4.2E-08	3.3E-08	2.8E-08	2.6E-08
Bladder Wall	1.2E-07	5.6E-08	4.2E-08	3.3E-08	2.8E-08	2.6E-08
Bone Surfaces	7.2E-06	1.7E-06	1.2E-06	1.4E-06	2.2E-06	7.4E-07
Brain	1.2E-07	5.5E-08	4.1E-08	3.3E-08	2.8E-08	2.6E-08
Breast	1.2E-07	5.5E-08	4.1E-08	3.3E-08	2.8E-08	2.6E-08
GI-Tract						
St Wall	1.4E-07	6.3E-08	4.5E-08	3.5E-08	2.9E-08	2.7E-08
SI Wall	1.5E-07	7.6E-08	5.2E-08	3.9E-08	3.1E-08	2.9E-08
ULI Wall	3.2E-07	1.9E-07	1.1E-07	7.2E-08	4.9E-08	4.3E-08
LLI Wall	7.1E-07	4.4E-07	2.3E-07	1.5E-07	9.3E-08	7.8E-08
Kidneys	2.6E-06	9.6E-07	5.9E-07	4.0E-07	3.0E-07	2.7E-07
Liver	5.4E-07	2.4E-07	1.7E-07	1.3E-07	1.0E-07	1.0E-07
Lungs	1.2E-07	5.5E-08	4.1E-08	3.3E-08	2.8E-08	2.6E-08
Muscle	1.2E-07	5.5E-08	4.1E-08	3.3E-08	2.8E-08	2.6E-08
Ovaries	1.3E-07	5.7E-08	4.3E-08	3.4E-08	2.8E-08	2.6E-08
Pancreas	1.2E-07	5.5E-08	4.2E-08	3.3E-08	2.8E-08	2.6E-08
Red Marrow	8.5E-07	1.9E-07	1.2E-07	1.2E-07	1.3E-07	7.6E-08
Skin	1.2E-07	5.5E-08	4.1E-08	3.3E-08	2.8E-08	2.6E-08
Spleen	1.2E-07	5.5E-08	4.1E-08	3.3E-08	2.8E-08	2.6E-08
Testes	1.4E-07	5.9E-08	4.5E-08	3.6E-08	2.8E-08	2.6E-08
Thymus	1.2E-07	5.5E-08	4.1E-08	3.3E-08	2.8E-08	2.6E-08
Thyroid	1.2E-07	5.5E-08	4.1E-08	3.3E-08	2.8E-08	2.6E-08
Uterus	1.2E-07	5.6E-08	4.2E-08	3.3E-08	2.8E-08	2.6E-08
Remainder	1.7E-07	7.1E-08	5.0E-08	3.7E-08	3.1E-08	2.8E-08
Effective Dose	3.5E-07	1.3E-07	8.5E-08	7.1E-08	7.0E-08	4.7E-08

GI-Tract Gastrointestinal Tract
St Stomach
SI Small Intestine
ULI Upper Large Intestine
LLI Lower Large Intestine

* In the biokinetic model for U parameter values for the adult apply to ages > 25 y. For radioisotopes of this element the dose coefficients for the adult are based on the 50-y integrated doses following an acute intake at age 25 y.

Table 5.-6.

Ingestion Dose Coefficients: Committed Equivalent and Effective Doses per Unit Intake (Sv/Bq) for U-236 (T1/2 = 2.3415E7 y)*

Age at intake	3 Months	1 Year	5 Years	10 Years	15 Years	Adult
Adrenals	1.3E-07	5.6E-08	4.2E-08	3.3E-08	2.8E-08	2.6E-08
Bladder Wall	1.3E-07	5.6E-08	4.2E-08	3.4E-08	2.8E-08	2.6E-08
Bone Surfaces	7.3E-06	1.7E-06	1.2E-06	1.4E-06	2.2E-06	7.4E-07
Brain	1.3E-07	5.6E-08	4.2E-08	3.3E-08	2.8E-08	2.6E-08
Breast	1.3E-07	5.6E-08	4.2E-08	3.3E-08	2.8E-08	2.6E-08
GI-Tract						
St Wall	1.4E-07	6.4E-08	4.6E-08	3.5E-08	3.0E-08	2.7E-08
SI Wall	1.5E-07	7.6E-08	5.2E-08	3.9E-08	3.1E-08	2.9E-08
ULI Wall	3.1E-07	1.8E-07	1.0E-07	6.9E-08	4.8E-08	4.2E-08
LLI Wall	6.4E-07	4.0E-07	2.1E-07	1.3E-07	8.5E-08	7.2E-08
Kidneys	2.6E-06	9.9E-07	6.0E-07	4.1E-07	3.1E-07	2.7E-07
Liver	5.5E-07	2.5E-07	1.7E-07	1.3E-07	1.1E-07	1.0E-07
Lungs	1.3E-07	5.6E-08	4.2E-08	3.3E-08	2.8E-08	2.6E-08
Muscle	1.3E-07	5.6E-08	4.2E-08	3.3E-08	2.8E-08	2.6E-08
Ovaries	1.3E-07	5.6E-08	4.2E-08	3.3E-08	2.8E-08	2.6E-08
Pancreas	1.3E-07	5.6E-08	4.2E-08	3.3E-08	2.8E-08	2.6E-08
Red Marrow	8.7E-07	1.9E-07	1.2E-07	1.2E-07	1.3E-07	7.7E-08
Skin	1.3E-07	5.6E-08	4.2E-08	3.3E-08	2.8E-08	2.6E-08
Spleen	1.3E-07	5.6E-08	4.2E-08	3.3E-08	2.8E-08	2.6E-08
Testes	1.3E-07	5.6E-08	4.2E-08	3.3E-08	2.8E-08	2.6E-08
Thymus	1.3E-07	5.6E-08	4.2E-08	3.3E-08	2.8E-08	2.6E-08
Thyroid	1.3E-07	5.6E-08	4.2E-08	3.3E-08	2.8E-08	2.6E-08
Uterus	1.3E-07	5.6E-08	4.2E-08	3.3E-08	2.8E-08	2.6E-08
Remainder	1.7E-07	7.3E-08	5.1E-08	3.8E-08	3.1E-08	2.9E-08
Effective Dose	3.5E-07	1.3E-07	8.4E-08	7.0E-08	7.1E-08	4.7E-08

GI-Tract Gastrointestinal Tract
St Stomach
SI Small Intestine
ULI Upper Large Intestine
LLI Lower Large Intestine

* In the biokinetic model for U parameter values for the adult apply to ages > 25 y. For radioisotopes of this element the dose coefficients for the adult are based on the 50-y integrated doses following an acute intake at age 25 y.

Table 5.-7.

Ingestion Dose Coefficients: Committed Equivalent and Effective Doses per Unit Intake (Sv/Bq) for U-238 (T1/2 = 4.468E9 y)*

Age at intake	3 Months	1 Year	5 Years	10 Years	15 Years	Adult
Adrenals	1.2E-07	5.3E-08	4.0E-08	3.1E-08	2.6E-08	2.5E-08
Bladder Wall	1.2E-07	5.3E-08	4.0E-08	3.1E-08	2.7E-08	2.5E-08
Bone Surfaces	6.9E-06	1.6E-06	1.2E-06	1.4E-06	2.1E-06	7.1E-07
Brain	1.2E-07	5.3E-08	4.0E-08	3.1E-08	2.6E-08	2.5E-08
Breast	1.2E-07	5.3E-08	4.0E-08	3.1E-08	2.6E-08	2.5E-08
GI-Tract						
St Wall	1.3E-07	6.0E-08	4.3E-08	3.3E-08	2.8E-08	2.6E-08
SI Wall	1.5E-07	7.1E-08	4.9E-08	3.7E-08	2.9E-08	2.7E-08
ULI Wall	2.9E-07	1.7E-07	9.6E-08	6.5E-08	4.5E-08	3.9E-08
LLI Wall	6.2E-07	3.8E-07	2.1E-07	1.3E-07	8.2E-08	6.9E-08
Kidneys	2.4E-06	9.2E-07	5.6E-07	3.9E-07	2.9E-07	2.5E-07
Liver	5.1E-07	2.3E-07	1.6E-07	1.2E-07	1.0E-07	9.6E-08
Lungs	1.2E-07	5.3E-08	4.0E-08	3.1E-08	2.6E-08	2.5E-08
Muscle	1.2E-07	5.3E-08	4.0E-08	3.1E-08	2.6E-08	2.5E-08
Ovaries	1.2E-07	5.4E-08	4.0E-08	3.2E-08	2.6E-08	2.5E-08
Pancreas	1.2E-07	5.3E-08	4.0E-08	3.1E-08	2.6E-08	2.5E-08
Red Marrow	8.3E-07	1.9E-07	1.2E-07	1.1E-07	1.3E-07	7.5E-08
Skin	1.2E-07	5.3E-08	4.0E-08	3.1E-08	2.6E-08	2.5E-08
Spleen	1.2E-07	5.3E-08	4.0E-08	3.1E-08	2.6E-08	2.5E-08
Testes	1.3E-07	5.6E-08	4.3E-08	3.5E-08	2.7E-08	2.5E-08
Thymus	1.2E-07	5.3E-08	4.0E-08	3.1E-08	2.6E-08	2.5E-08
Thyroid	1.2E-07	5.3E-08	4.0E-08	3.1E-08	2.6E-08	2.5E-08
Uterus	1.2E-07	5.3E-08	4.0E-08	3.1E-08	2.6E-08	2.5E-08
Remainder	1.6E-07	6.8E-08	4.8E-08	3.6E-08	2.9E-08	2.7E-08
Effective Dose	3.3E-07	1.2E-07	8.0E-08	6.8E-08	6.7E-08	4.5E-08

GI-Tract Gastrointestinal Tract
St Stomach
SI Small Intestine
ULI Upper Large Intestine
LLI Lower Large Intestine

* In the biokinetic model for U parameter values for the adult apply to ages > 25 y. For radioisotopes of this element the dose coefficients for the adult are based on the 50-y integrated doses following an acute intake at age 25 y.

References

Bassett, S. H., Frenkel, A., Cedars, N., VanAlstine, H., Waterhouse, C. and Cusson, K. (1948). *The Excretion of Hexavalant Uranium following Intravenous Administration. II. Studies on Human Subjects*, pp. 1–57. Rochester, NY: University of Rochester.

Bernard, S. R. and Struxness, E. G. (1957). *A Study of the Distribution and Excretion of Uranium in Man.* Oak Ridge National Laboratory, Oak Ridge, TN, ORNL-2304.

Bernard, S. R., Muir, J. R. and Royster, G. W. (1957). The distribution and excretion of uranium in man. In: *Proceedings of the Health Physics Society First Annual Meeting*, pp. 33–48.

Bhattacharyya, M. H., Larsen, R. P., Cohen, N., Ralston, L. G., Moretti, E. S., Oldham, R. D. and Ayres, L. (1989). Gastrointestinal absorption of plutonium and uranium in fed and fasted adult baboons and mice: Application to humans. *Radiat. Prot. Dosim.* **26**, 159–165.

Brambell, F. M. R. (1970). Transmission of passive immunity from mother to young. Frontiers. In: *Biology*, Vol. 18, Amsterdam, North Holland.

Broadway, J. A. and Strong, A. B. (1983). Radionuclides in human bone samples. *Health Phys.* **45**, 765–768.

Campbell, E. E., McInroy, J. F. and Schulte, H. F. (1975). Uranium in the tissue of occupationally exposed workers. In: *Conference on Occupational Health: Experience with Uranium*, ERDA 93, pp. 324–349. Arlington, VA, April 28–30, 1975.

Chevari, S. and Likhner, D. (1968). Complex formation of natural uranium in the blood. *Med. Radiol.* **13**, 53–57 (Russian), ANL-tr- 898.

Cooper, J. R., Stradling, G. N., Smith, H. and Ham, S. E. (1982). The behaviour of uranium-233 oxide and uranyl-233 nitrate in rats. *Int. J. Radiat. Biol.* **41**, 421–433.

Donoghue, J. K., Dyson, E. D., Hislop, J. S., Leach, A. M. and Spoor, N. L. (1972). Human exposure to natural uranium: A case history and analytical results from some *postmortem* tissues. *Br. J. Industr. Med.* **29**, 81–89.

Durbin, P. W. (1984). Metabolic models for uranium. In: *Biokinetics and Analysis of Uranium in Man* (Moore, R. H. ed.). United States Uranium Registry, USUR-05, HEHF-47, 1984:F1-F65. Available from National Technical Information Service, 5285 Port Royal Road, Springfield, VA, 22151.

Fisenne, I. M. and Perry, P. M. (1985). Isotopic U concentration in human blood from New York City donors. *Health Phys.* **49**, 1272–1275.

Fisenne, I. M. and Welford, G. A. (1986). Natural U concentrations in soft tissues and bone of New York City residents. *Health Phys.* **50**, 739–746.

Fisenne, I. M., Perry, P. M. and Harley, N. H. (1988). Uranium in humans. *Radiat. Prot. Dosim.* **24**, 127–131.

Gonzales, E. R. and McInroy, J. F. (1991). The distribution of uranium in two whole bodies. Abstracts of papers presented at the 36th annual meeting of the Health Physics Society. *Health Phys.* **60** (Suppl. 2), 51.

Harduin, J. C., Royer, Ph. and Piechowski, J. (1994). *Uptake and Urinary Excretion of Uranium after Oral Administration in Man.* Proc. *Radiat. Prot. Dosim.* **53**(1–4), 245–248.

Harrison, J. D. (1991). The gastrointestinal absorption of the actinide elements. *Sci. Total Environ.* **100**, 43–60.

Hursh, J. B. and Spoor, N. L. (1973). Data in man. In: *Uranium, Plutonium and Transplutonium Elements*, 197–239 (Hodge, H. C., Stannard, J. N. and Hursh, J. B. eds) Springer, Berlin.

Hursh, J. B., Neuman, W. R., Toribara, T., Wilson, H. and Waterhouse, C. (1969). Oral ingestion of uranium by man. *Health Phys.* **17**, 619–621.

ICRP (1979). *Limits for Intakes of Radionuclides by Workers.* ICRP Publication 30, Part 1. *Annals of the ICRP* **2**(3/4), Pergamon Press, Oxford.

ICRP (1988). *Individual Monitoring for Intakes of Radionuclides by Workers: Design and Interpretation,* ICRP Publication 54. *Annals of the ICRP* **19**(1–3), Pergamon Press, Oxford.

ICRP (1989). *Age-dependent Doses to Members of the Public from Intake of Radionuclides: Part 1,* ICRP Publication 56. *Annals of the ICRP* **20**(2), Pergamon Press, Oxford.

ICRP (1993). *Age-dependent Doses to Members of the Public from Intake of Radionuclides: Part 2 Ingestion Dose Coefficient,* ICRP Publication 67. *Annals of the ICRP* **23**(3/4), Elsevier Science Ltd, Oxford.

Igarashi, Y., Yamakawa, A., Seki, R. and Ikeda, N. (1985). Determination of U in Japanese human tissues by the fission track method. *Health Phys.* **49**, 707–712.

Jones, E. S. (1966). Microscopic and autoradiographic studies of distribution of uranium in the rat kidney. *Health Phys.* **12**, 1437–1451.

Kathren, R. L., McInroy, J. F., Moore, R. H. and Dietert, S. E. (1989). Uranium in the tissues of an occupationally exposed individual. *Health Phys.* **57**, 17–21.

Kisieleski, W., Faraghan, W. G., Norris, W. P. and Arnold, J. S. (1952). The metabolism of uranium-233 in mice. *J. Pharmacol. Exp. Ther.* **104**, 459–467.

Larsen, R. P. and Orlandini, K. A. (1983). Gastrointestinal absorption of uranium in man. In: *Argonne National Laboratory, Environmental Research Division Annual Report*, 46–50, July 1982–June 1983. ANL-83-100-Pt.2, Argonne, IL.

Leggett, R. W. (1989). The behaviour and chemical toxicity of uranium in the kidney: A reassessment. *Health Phys.* **57**, 365–383.

Leggett, R. W. (1992). A generic age-specific biokinetic model for calcium-like elements. *Radiat. Protect. Dosim.* **41**, 183–198.

Leggett, R. W. and Harrison, J. D. (1995). Fractional absorption of ingested uranium in humans. *Health Phys.* **68**, 484–498.

Lianquing, L. and Guiyun, L. (1990). Uranium concentration in bone of Beijing (China) residents. *Sci. Total Environ.* **90**, 267–272.

Limson-Zamora, M. L., Zielinski, J. and Falcomer, R. A. (1992). A study of the human urinary excretion of chronically ingested uranium — insights on the gastrointestinal absorption factor. In: *Proceedings of the 8th International Congress of IRPA*, Vol. II, pp. 1085–1088.

Lipsztein, J. L. (1981). *An Improved Model for Uranium Metabolism in the Primate,* Ph.D. Dissertation, New York University.

Lucas, H. F. and Markun, F. (1970). Thorium and uranium in blood, urine, and cigarettes. In: *Radiological Physics Division Annual Report,* 47–52, July 1969–June 1970, Argonne National Laboratory, ANL-7615.

Luessenhop, A. J., Gallimore, J. C., Sweet, W. H., Struxness, E. G. and Robinson, J. (1958). The toxicity in man of hexavelant uranium following intravenous administration. *Am. J. Roentgenol.* **79**, 83–100.

Morrow, P. E., Gelein, R. M., Beiter, H. D., Scott, J. B., Picano, J. J. and Yuile, C. L. (1982). Inhalation and intravenous studies of UF_6/UO_2F_2 in dogs. *Health Phys.* **43**, 859–873.

Muir, J. R., Fish, B. R., Jones, E. S., Gillum, N. L. and Thompson, J. L. (1960). Distribution and excretion of uranium. In: *Health Physics Division Annual Progress Report for Period Ending July 31, 1960,* pp. 272–273. Oak Ridge National Laboratory. Oak Ridge, TN, ORNL-2994.

NEA/OECD (1988). *Committee on Radiation Protection and Public Health.* Report of an Expert Group on Gut Transfer Factors. NEA/OECD, Paris.

Neuman, W. F., Fleming, R. W., Dounce, A. L., Carlson, A. B., O'Leary, J. and Mulryan, B. (1948). The distribution and excretion of injected uranium. *J. Biol. Chem.* **173**, 737–748.

Priest, N. D., Howells, G. R., Green, D. and Haines, J. W. (1982). Uranium in bone: Metabolic and autoradiographic studies in the rat. *Human Toxicol.* **1**, 97–114.

Roberts, A. M., Coulston, D. J. and Bates, T. H. (1977). Confirmation of *in vivo* uranium-in-chest survey by analysis of autopsy specimens. *Health Phys.* **32**, 435–437.

Rowland, R. E. and Farnham, J. E. (1969). The deposition of uranium in bone. *Health Phys.* **17**, 139–144.

Singh, N. P., Lewis, L. L. and Wrenn, M. E. (1986). Uranium in human tissues of Colorado, Pennsylvania and Utah populations. In: *Thirty-first Annual Meeting of the Health Physics Society,* Vol. 50, p. S83 (abstract), June 29–July 3, 1986, Pittsburgh, Pennsylvania. New York: Pergamon Press.

Singh, N. P., Bennett, D. B., Wrenn, M. E. and Saccomanno, G. (1987). Concentrations of alpha-emitting isotopes of U and Th in uranium miners' and millers' tissues. *Health Phys.* **53**, 261–265.

Sontag, W. (1984). Long-term behaviour of Pu-239, Am-241 and U-233 in different bones of one-year-old rats: Macrodistribution and macrodosimetry. *Human Toxicol.* **3**, 469–483.

Spencer, H., Osis, D., Fisenne, I. M., Perry, P. M. and Harley, N. H. (1990). Measured intake and excretion patterns of naturally occurring ^{234}U, ^{238}U, and calcium in humans. *Rad. Res.* **124**, 90–95.

Stevens, W., Bruenger, F. W., Atherton, D. R., Smith, J. M. and Taylor, G. N. (1980). The distribution and retention of hexavelant ^{233}U in the beagle. *Radiat. Res.* **83**, 109–126.

Struxness, E. G., Luessenhop, A. J., Bernard, S. R. and Gallimore, J. C. (1956). The distribution and excretion of hexavelant uranium in man. In: *Proceedings of International Conference on the Peaceful Uses of Atomic Energy,* Vol. 10, pp. 186–195. New York, United Nations.

Sullivan, M. F. (1980). Absorption of actinide elements from the gastrointestinal tract of neonatal animals. *Health Phys.* **38**, 173–185.

Sullivan, M. F. and Gorham, L. S. (1982). Further studies on the absorption of actinide elements from the gastrointestinal tract of neonatal animals. *Health Phys.* **43**, 509–519.

Svyatkina, N. S. and Novikov, Yu. V. (1975). The problem of absorption of natural uranium in the gastrointestinal tract of man. *Gig. Sanit. J.* **43**(1), 43–45.

Terepka, A. R., Toribara, T. Y. and Neuman, W. F. (1964). *Skeletal Retention of Uranium in Man.* Abstract 22, 46th meeting of the endocrine society, San Francisco, CA, (see pp. 206–207 of Hursh and Spoor, 1973).

Wrenn, M. E., Durbin, P. W., Howard, B., Lipzstein, J. L., Rundo, J., Still, E. T. and Willis, D. L. (1985). Metabolism of Ingested U and Ra. *Health Phys.* **48**, 601–603.

Wrenn, M. E., Singh, N. P., Ruth, H., Rallison, M. I. and Burleigh, D. P. (1989). Gastrointestinal absorption of soluble uranium from drinking water by man. *Radiat. Prot. Dosim.* **26**, 119–122.

Wrenn, M. E., Bertelli, L., Durbin, P. W., Singh, N. P., Lipsztein, J. L. and Eckerman, K. F. (1994). A comprehensive metabolic model for uranium metabolism and dosimetry based on human and animal data. *Radiat. Prot. Dosim.* **53**(1–4), 255–258.